DATE DUE

GAYLORD   PRINTED IN U.S.A.

Is Mathematics Inevitable?

A Miscellany

Copyright ©2008 by
The Mathematical Association of America

ISBN 978-0-88385-566-9

Library of Congress number 2007940798

Printed in the United States of America

Current printing (last digit):
10 9 8 7 6 5 4 3 2 1

Is Mathematics Inevitable?

A Miscellany

edited by

Underwood Dudley
Professor Emeritus, DePauw University

Published and Distributed by
The Mathematical Association of America

SPECTRUM SERIES

The Spectrum Series of the Mathematical Association of America was so named to reflect its purpose: to publish a broad range of books including biographies, accessible expositions of old or new mathematical ideas, reprints and revisions of excellent out-of-print books, popular works, and other monographs of high interest that will appeal to a broad range of readers, including students and teachers of mathematics, mathematical amateurs, and researchers.

MAA Service Center
P.O. Box 91112
Washington, DC 20090-1112
800-331-1622   FAX 301-206-9789

Short Preface

A traditional function of a preface is to give excuses for the book that follows it, no matter how inexcusable it is, and, since mathematics is a subject with a long tradition, it would not do to depart from custom.

I should first apologize for the disingenuous title. It was meant to catch your eye and, if you are reading this, it has succeeded. It does not, however, describe the contents of the book, which is a collection of more or less unrelated pieces. One of them considers the question of the title and gives the answer, "yes." A more accurate title for the book, though less appealing, would be "Some Mathematical Stuff."

I think that the stuff that follows has some interesting things in it that the reader would probably not encounter otherwise. The book is not an assemblage of Mathematics' Greatest Hits, nor of classics. It does not have a theme.

Nevertheless, I think that it deserves to exist. Writing on mathematics has more of a claim to be preserved and reprinted than does most writing. What appears in newspapers is notoriously ephemeral, and magazines are much the same. Not many people would want to read issues of, say, *Popular Science* in the 1950s for their content. Even books can quickly become dated and irrelevant. Mathematics, however, is permanent. There are gems to be found in the literature of mathematics, periodical or otherwise, that shine as brightly today as when they first appeared in print, and they deserve to be seen and admired.

This is not to say that everything that follows is a flawless jewel, nor that there are not other items that deserve equal or greater exposure. I have been buying mathematics books and subscribing to mathematics journals (and even reading some of them) for many decades and what is included here was, for the most part, taken from my shelves. I hope the reader will find some of it interesting, entertaining, enlightening, or all three at once.

That's probably enough excuses.

Underwood Dudley

Contents

1
Dieudonné on Mathematics

The French pride themselves on their precision and clarity of thought, and it is possible that they are justified in doing so. There follows a selection by Jean Dieudonné, one of the founders of the Bourbaki group and one of the powerful mathematicians of the twentieth century, written for non-mathematicians. It is indeed precise and clear and it is well worth reading.

Some of what he says may not be entirely good for the egos of those of us in the mathematics business. For example, he defines "mathematician" as "someone who has published the proof of at least one non-trivial theorem." He also asserts that there are in the United States only about 600 mathematicians "who can usefully take on teaching of the kind needed for graduate courses through which new ideas are spread, and who can effectively advise young mathematicians engaged in research." Dieudonné was an Olympian and hence entitled to god-like views, but I would place the bar lower. Surely a PhD degree in mathematics is enough?

Mathematics and Mathematicians

Jean Dieudonné (1992)

1. The Concept of Mathematics

The position of mathematics among human activities is paradoxical. In our day almost everybody in the "developed" countries knows that it is an important discipline, needed for most branches of science and technology, and that an ever greater number of professions cannot be exercised without some knowledge of it. Nevertheless, if you ask "What is mathematics?" or "What does a mathematician do?", you will very rarely get anything but a preposterous answer, unless your interlocutor has had a mathematical training at least up to the end of the first two years at a university. Even men who are eminent in other sciences often have only aberrant notions about the work of mathematicians.

Since everybody has made contact with mathematics by means of numerical calculations at primary school, the most widespread idea is that a mathematician is someone who is a virtuoso in these calculations. With the advent of computers and their language, people will now think that that means somebody who is especially good at "programming" them, and who spends all his time doing this. Engineers, always looking for optimal values for the measures of magnitudes which interest them, think of mathematicians as custodians of a fund of formulae, to be supplied to them on demand. But the most fallacious of current conceptions, firmly held by nearly all our contemporaries, inundated as they are by plentiful descriptions in the media of the progress made in all other sciences, is that there is nothing more to be discovered in mathematics, and that the work of mathematicians is limited to passing on the legacy of past centuries.

Must we rest content with this state of affairs, and do no more than remark, with A. Weil that "mathematics has this peculiarity, that it is not understood by non-mathematicians"? I think that we must at least try to find the reasons for this lack of understanding. There are in fact numerous periodicals devoted to popularizing recent scientific discoveries, addressed to large numbers of readers and at all levels of scientific education. With a few exceptions, there is nothing of the kind on recent progress in mathematics (which could make people think that none has taken place!); by contrast, there is plenty of literature on astrophysics, geology, chemistry, molecular biology, even atomic or nuclear physics. It seems that it is possible, with the help of diagrams in which details are eliminated and the essential objects clearly displayed, in addition to explanations which simplify radically what specialists have been able to deduce from complex and delicate experiments, to give readers of these publications the illusion that they understand what an atom, or a gene, or a galaxy is, even if the orders of magnitude concerned, ranging from an angstrom (10^{-10} meters) to a light year (3×10^{15} meters), are beyond the grasp of our imaginations.

By way of contrast, let us take one of the most fruitful theories of modern mathematics, the one known as "sheaf cohomology". Thought of in 1946, it is more or less contemporary with the "double helix" of molecular biology, and has led to progress of comparable magnitude. I should, however, be quite unable to explain in what this theory consists to someone who had not followed at least the first two years of a course in mathematics at a university. Even for an able student at this level the explanation would take several hours; while to explain the way in which the theory is used would take a good deal more time still. This is because here we can no longer make use of explanatory diagrams; before we can get to the theory in question we must have absorbed a dozen notions equally abstract: topologies, rings, modules, homomorphisms, etc., none of which can be rendered in a "visual" way.

The same remarks can be applied to almost all ideas which form the basis for the great mathematical theories of our day. I must therefore resign myself: if I want to discuss what is accessible at baccalaureate level, I can allude only to elementary concepts under three headings: algebra, number theory and set theory; adding from time to time a remark addressed to readers who have some familiarity with the calculus. Beginning with these notions, I shall try to explain how they have evolved of necessity, revealing underlying notions which are much more abstract, and acquiring in the process an incomparably greater efficacy in solving mathematical problems. To go further would mean putting together streams of words in which the non-mathematical reader would very soon lose his footing. The list I have mentioned is there only to give an idea of the minute number of theorems which I shall discuss, compared with the immensity of modern mathematical knowledge.

2. A Mathematician's Life

The term "musician", in common speech, can mean a composer, a performer or a teacher of music, or any combination of these activities. In the same way, by "mathematician" we can understand a teacher of mathematics, one who uses mathematics, or a creative mathematician; and the common belief is that this last species is now extinct. Since, however, it is my intention to help my readers to understand the origin and nature of modern mathematics (of which the greater part dates from no earlier than 1840), we shall be dealing almost exclusively with this last category. A mathematician, then, will be defined in what follows as someone *who has published the proof of at least one non-trivial theorem* (the meaning of the expression "trivial" will be precisely defined later).

It seems to me that it would be not uninteresting for the reader to be taken on a little trip to the unknown country of mathematicians before we try to explain what they do.

There is every reason to think that an aptitude for creating mathematics is independent of race, and it does not seem on the whole to be transmitted by heredity: mathematicians who are the offspring of mathematicians have always been rare exceptions. But in the flowering of a mathematical talent social environment has an important part to play. There is no example, even among the greatest geniuses, of a mathematician discovering for himself, *ab ovo*, the mathematics known to his generation, and the story told of Pascal relating to this is only a legend. The social milieu, then, must be such as to enable the future mathematician to receive at least some elementary instruction exposing him to genuine proofs, awakening his curiosity, and after that enabling him to come into contact with the mathematics of his time: something which depends on his having access to the works where it is written down. Up until the end of the eighteenth century, there was no real organized university education in mathematics, and, from Descartes and Fermat to Gauss and Dirichlet, nearly all the great mathematicians trained themselves without instructors, by reading the works of their illustrious predecessors (still an excellent exercise!). Even today it is likely that many mathematical talents never come to light, for lack of a favorable social environment, and it is not surprising that we do not know of mathematicians in less "developed" societies. Even in countries which have evolved further, the type of elementary instruction can be unpropitious for the flowering of mathematical vocations when it is subject to religious or political constraints, or to exclusively utilitarian concerns, as was the case in the United States until well into the twentieth century. Secondary education there is in effect in the hands of local authorities, who often used to think it was more important for adolescents to learn typing or how to drive a car than Latin or Euclidean geometry. I have known famous American mathematicians who were born in small towns and had never seen a mathematical proof until they were 18 years old; attracted to science, they were oriented at the university toward a career in engineering, and it was only by chance that they were confronted with real mathematical instruction, and thus discovered their true vocations.

In societies where education is available to a sufficiently wide section of society (if necessary with the help of grants), the social origins of mathematicians are very varied. They may come from the nobility, like Fagnano, Riccati and d'Alembert, from the upper middle class, like Descartes, Fermat, Pascal, Kronecker, Jordan, Poincaré and von Neumann, or they may come from very humble backgrounds, like Roberval, Gauss, Élie Cartan and Lebesgue. Most are born into middle-class families, often quite penurious ones.

A mathematical vocation is most often awakened at about fifteen years of age, but it can be held back by an education which does not allow for any concept of proof, as was the case in the United States, as stated above. In any case, contrary to a fairly widespread opinion, the creative period rarely begins before the age of twenty-three to twenty-five; the cases of Pascal, Clairaut, Lagrange, Gauss, Galois, Minkowski, and, in our time, Deligne, who were responsible for notable mathematical discoveries at less than twenty years old, are exceptional. As for mathematical longevity, another current opinion is that, as Hardy said, "mathematics is a young man's game". It is true that a number of major discoveries have been made by mathematicians who were not much older than thirty. Still, for many of the greatest, such as Poincaré, Hilbert or H. Weyl, the creative period is prolonged, and they remain quite productive up to the age of fifty or fifty-five. Others, such as Killing, Emmy Noether, Hardy (as he himself recognized) or more recently Zariski and Chevalley, give of their best when they are about 40. Finally, a few mathematicians favored by the gods, like Kronecker, Elie Cartan, Siegel, A. Weil, J. Leray, and I. Gelfand, have continued to prove fine theorems after the age of sixty. All the same, just as in many sports one can hardly expect to remain a champion after thirty, so most mathematicians have to resign themselves to seeing their creative imaginations become barren after the age of fifty-five or sixty. For some it is a heartbreaking experience, as Hardy's *Apology* movingly witnesses; others adapt by finding an activity in which they can exercise those faculties which they still possess.

As with many scholars, the life of a mathematician is dominated by an insatiable curiosity, a desire bordering on passion to solve the problems he is studying, which can cut him off completely from the realities around him. The absent-mindedness or eccentricity of famous mathematicians comes simply from this. The fact is that the discovery of a mathematical proof is usually reached only after periods of intense and sustained concentration, sometimes renewed at intervals over months or years before the looked-for answer is found. Gauss himself acknowledged that he spent several years seeking the sign of an algebraic equation, and Poincaré, on being asked how he made his discoveries, answered "by thinking about them often"; he has also described in detail the unfolding of the course of reflections and attempts which led to one of his finest results, the discovery of "Fuchsian" functions.

What a mathematician seeks above all, then, is to have at his disposal enough time to devote himself to his work, and that is why, since the nineteenth century, they have preferred careers in teaching in universities or colleges, where the hours given to teaching are relatively few and the holidays are long. Remuneration is of only secondary importance, and we have recently seen, in the United States among other places, mathematicians abandoning lucrative posts in industry in order to return to a university, at the cost of a considerable drop in salary.

Moreover it is only recently that the number of teaching posts in universities has become fairly large. Before 1940 it was very restricted (less than one hundred in France), and up until 1920 mathematicians of the quality of Kummer, Weierstrass, Grassmann, Killing and Montel had to rest content with being teachers in secondary education during all or part of their working lives; a situation which persisted for a long time in small countries with few universities.

Before the nineteenth century, a mathematician's opportunities to do his work were even more precarious, and failing a personal fortune, a Maecaenas or an academy to provide him with a decent living, there was almost no other resource than to be an astronomer or a surveyor (as in the case of Gauss, who devoted a considerable amount of time to these professions). It is this circumstance which no doubt explains the very restricted number of talented mathematicians before 1800.

This same need to devote long hours of reflection to the problems which they are trying to solve almost automatically precludes the possibility of driving in tandem absorbing tasks in other areas of life (such as administration) and serious scientific work. The case of Fourier, Prefect of Isère at the time when he was creating the theory of heat, is probably unique. Where we find mathematicians occupying high positions in administration or in government, they will have essentially given up their research during the time that they are discharging these functions. Besides which, as Hardy commented ironically, "the later records of mathematicians who have left mathematics are not particularly encouraging".

As regards the personal traits of mathematicians, if we except their absent-mindedness and the lack of "practical know-how" with which many are afflicted, it can be said that there is hardly any difference between them and any other section of the population of comparable status taken at random. The same variety in their characters is to be found, the same virtues and vices. If they have at times a tendency to pride themselves on the possession of a rather uncommon talent, they must remember that many other techniques can be performed by only a small number of people. No doubt there are far fewer tightrope walkers who can dance on the high wire, or chess players who can play blindfolded. Nevertheless, as Hardy says, what a mathematician does "has a certain character of permanence", and you have not lived in vain if you have added a small stone to the pyramid of human knowledge.

In any case, mathematicians must accept that their talent does not confer on them any particular competence outside their own domain, as those who have thought that their prestige among their colleagues would enable them to reform society have learnt to their cost.

The opinions of mathematicians in matters of religion or politics are moreover very diverse: Cauchy was a bigot, but Hardy was a curious kind of atheist for whom God was a personal enemy; Gauss was very conservative, but Galois was an ardent revolutionary. An extreme and rather distressing case is that of the young German mathematician O. Teichmüller, who perhaps had a genius comparable to that of Galois; but he was a fanatical Nazi, who contributed to the expulsion of his Jewish teachers. He went to the Russian front as an SS officer, and did not return.

Most mathematicians lead the lives of good citizens, caring little about the tumults which convulse the world, desiring neither power nor wealth, and content with modest comfort. Without actively seeking honors, most are not indifferent to them. In contrast to those of many artists, their lives are rarely thrown into chaos by emotional storms; lovers of sensation and romance have little to glean among them, and have to content themselves with more or less romanticized biographies of Galois, of Sonia Kowalewska or of Ramanujan.

3. The Work of Mathematicians and the Mathematical Community

Research in the experimental sciences is done in laboratories, where larger and larger teams are needed to manipulate the instruments and to scrutinize the results. To do research in mathematics nothing is needed except paper and a good library. Teamwork, as practiced in the experimental sciences is, then, quite unusual in mathematics, most mathematicians finding it difficult to think seriously except in silence and solitude. Collaborative work, while quite common, most often consists in putting together results that each of the collaborators has managed to obtain in isolation, albeit with mutual profit from each other's ideas, enabling them to progress from new points of departure. The best known example of prolonged collaboration between mathematicians is that of Hardy and Littlewood: one of them lived in Oxford and the other in Cambridge, and they saw each other only occasionally; their work together being done entirely through correspondence.

But if most publications are the work of individuals, those mathematicians are rare who, like Grassmann, Hensel or Élie Cartan, can work productively for a long time in almost complete isolation (due, in these three cases to the novelty of their ideas, which were not understood by their contemporaries). Most become discouraged if they lack the opportunity to communicate quite frequently with their colleagues in the hope that they will understand them, the more so because the abstract nature of their research makes it difficult to exchange ideas with non-mathematicians.

Until the middle of the seventeenth century the only means of communication was private correspondence (sometimes centralized by well-disposed letter-writers, such as Mersenne or Collins), personal visits, and sometimes the printing of a work at the author's expense (unless some Maecaenas consented to pay). The first publications to be regularly issued by academies appeared about 1660, but the numbers were still small, and the few scientific journals which were published before 1820 were devoted to science in general, without specialization. The first specialist mathematical journals to cross national boundaries were the *Journal de Crelle*, founded in 1826, followed closely by the *Journal de Liouville* in 1835.

It was now, however, that linguistic barriers began to arise. In the seventeenth century almost all scientific publications were produced in Latin (Descartes' *Géométrie*, with some of Pascal's writings, being the only noteworthy exceptions). This tradition began to decline in the eighteenth century; French mathematicians now wrote only in their own language, and abroad some works of an instructional nature were written in their authors' languages. After Gauss's time, the use of Latin rapidly disappeared in scientific publications; and this continued to slow down the dissemination of new results, especially in France, where there was hardly any study of living languages until the end of the nineteenth century. A rather ridiculous example occurred when the Paris Academy of Science, in 1880, set up a competition to solve a problem in number theory which had already been solved more than 20 years before by H. J. S. Smith. Apparently Hermite and Jordan did not read English. In our day, as is well known, English is in a fair way to becoming a universal language of science.

Throughout the course of the nineteenth century the number of mathematical journals continued to grow steadily, a process which was accelerated after 1920 with the increase in the number of countries where mathematical studies were being developed, and which culminated, after 1950, in a veritable explosion; so that today there are about 500 mathematical periodicals in the world. To prevent people from drowning in this ocean of information, journals devoted exclusively to listing and summarizing other publications have been created since the last third of the nineteenth century. The most widely disseminated of these journals, the American *Mathematical Reviews*, now extends to almost 4,000 pages a year (with an average of 5 to 10 articles analyzed on each page).

Along with the multiplication of periodicals has gone that of expository works, often grouped together in series of monographs (at times quite specialized). Thus it is rare nowadays for a new theory to wait more than 10 years before becoming the subject of instructional expositions.

Toward the middle of the nineteenth century, there was born in German universities the practice of holding "seminars": under the direction of one or more teachers, several mathematicians, local or visiting, among whom are often included students working for doctorates, analyze the state of a problem, or review the most noted new results in the course of periodic sessions during the academic year. Since 1920, this practice has spread all over the world; the expositions which are made at seminars are often disseminated in the form of printed papers, and can thus reach a larger public. The same thing happens with specialized courses taught in many universities and addressed to advanced students.

It has, however, long been evident that in the field of science verbal exchanges are often more productive than the reading of papers. It has been traditional, since the Middle Ages, for students to travel from one university to another, and this tradition is perpetuated in our day, notably in Germany. In addition, invitations to academics to come to work and teach in universities other than their own have become quite frequent.

This need for personal contacts has been institutionalized since 1897 in the International Congress of Mathematicians, which, since 1900, has met every four years (with two interruptions caused by the world wars). The ever larger numbers participating in these Congresses have to some extent reduced their usefulness, and since 1935 more limited gatherings have proliferated: colloquia, symposia, workshops, summer schools, etc., where a more-or-less open-ended range of specialists meet to take stock of their recent discoveries and discuss the problems of the day.

4. Masters and Schools

The spirit of competition, the concern with ranking, the race for records and prizes have never been so furious as in our day: this fact stares you in the face where sport is concerned, and the same spirit extends to intellectual pursuits such as chess or bridge, and even to many others more or less laughable.

In spite of this, we have recently seen passionately advocated the view that all human beings have the same capacity for intellectual creativity, and that the flagrant inequalities we notice in this respect depend only on whether individuals are more or less favored by education. A curious doctrine, which entails the belief that the brain has a different physiology from other organs, and at which we can only shrug our shoulders when we think of all the children of princes or of millionaires who have remained incurably stupid despite all the care taken to provide them with the best instructors.

We must then admit that, as in all disciplines, ability among mathematicians varies considerably from one individual to another. In the "developed" countries, the teaching of mathematics is assured by a large staff of academics, most of whom have had to obtain a doctorate (or its equivalent) by producing a work of original research, but these are far from being all equally gifted in research capability, since they may lack creative imagination.

The great majority of these works are in fact what we call trivial, that is, they are limited to drawing some obvious conclusions from well-known principles. Often the theses of these mathematicians are not even published; inspired by a "supervisor", they reflect the latter's ideas more than those of their authors. And so, once left to their own resources, such people publish articles at rare intervals relating directly to their theses and then very soon cease all original publication. All the same, these people have an indubitable importance: apart from the dominant social role which they play in the education of all the scientific elite of a country, it is they who can pick out, in their first years at the university, the particularly gifted students who will be the mathematicians of the next generation. If they are able to keep up to date with advances in their science and to use the knowledge to enrich their teaching, they will awaken hesitant vocations and will direct them to colleagues charged with guiding the first steps of future researchers.

At a higher level we find a far less numerous category (especially in countries like the United States, which need huge teaching staffs), the class of mathematicians capable of going beyond their thesis work, and even of becoming involved in totally different areas; they often remain active in research for thirty years or so, and publish several dozen original papers. In the "developed" countries you can say that, taking the good years with the bad, such a mathematician is born each year for every ten million people, which gives you about 150 active creative mathematicians in a country the size of France, and 600 in the United States or the Soviet Union. It is only they who can usefully take on teaching of the kind needed for graduate courses through which new ideas are spread, and who can effectively advise young mathematicians engaged in research.

Finally, there are the great innovators, whose ideas divert the entire course of the science of their time and which sometimes have repercussions over more than a century. But, as Einstein said to Paul Valéry, "an idea is so rare!" (obviously Einstein was thinking of a *great* idea). We can count half a dozen geniuses of this kind in the eighteenth century, thirty or so in the nineteenth, and nowadays we can reckon that one or two a year arise in the whole world. To work with these great mathematicians is an uplifting and enriching experience;

Hardy saw in his association with Littlewood and Ramanujan "the decisive event of his life", and I can say as much of my collaboration with the Bourbaki team.

There is no Nobel Prize for mathematicians. But since 1936 the International Congress of Mathematicians, which meets every four years, has awarded at each of its sessions two, three or four medals, known as "Fields medals", to mathematicians, preferably less than 40 years old, whose work is judged the most outstanding by an international committee. Up to now there have been ten Fields medalists from the United States, five French, four English, three Scandinavian, two Russian, two Japanese, one German, one Belgian, one Chinese and one Italian. It is always possible to criticize the way in which these prizes are distributed, or even their principle. I think, however, that nobody can dispute the merit of the medalists, all of whom have had at least one great idea in the course of their career. We should, moreover, add to this list some fifteen mathematicians of comparable excellence but who could not be rewarded on account of the restricted number of medals.

The question of nationality has no greater importance for mathematicians than for research-workers in other sciences; once the linguistic barriers have been overcome (most often through the use of English), a French mathematician will feel much closer, as regards his concepts and his methods, to a Chinese mathematician than to an engineer from his own country.

Natural associations between mathematicians are based much less on countries than on schools. While many of these latter are centered on a country, there are countries where several schools flourish, and schools whose influence extends well beyond national frontiers. These are not, of course, entities clearly defined in time and space, but groupings attached to a continuing tradition, with masters in common, and certain preferred subjects and methods.

Before 1800, mathematicians, few in number and scattered, did not have pupils, properly speaking, even though, from the middle of the seventeenth century, communications between them were frequent, and research in mathematics saw almost uninterrupted progress. The first school of mathematics, chronologically speaking, was formed in Paris after the Revolution, thanks to the foundation of the École Polytechnique, which provided a nursery for mathematicians until about 1880, after which date the baton was taken up by the École Normale Supérieure. In Germany, Gauss was still isolated, but the generations which followed created centers of mathematical research in several universities, of which the most important were Berlin and Göttingen. In England, mathematical research in universities entered a phase of lethargy after 1780; it did not wake up until about 1830, with the Cambridge school, which was especially prolific during the nineteenth century in logic and in algebra, and has continued to shine in mathematical physics up to our own day. Italy had only a few isolated mathematicians from 1700 to 1850, but several active schools have developed there since, notably in algebraic geometry, differential geometry and functional analysis.

During the twentieth century, owing to wars and revolutions, schools of mathematics have experienced many vicissitudes. There is first a phenomenon which is difficult to explain in view of the very different social and political conditions in the countries where it occurred. After the 1914-1918 war there suddenly appeared a constellation of mathematicians of the first order in the Soviet Union and in Poland, both of which had up to then produced only a small number of scientists of international renown. The same phenomenon occurred in Japan after the 1939-1945 war, even though the poor and rigid university system there meant that the Japanese school lost some of its best members to the United States.

The factors which can impede mathematical progress are easier to understand. In France the body of young scientists was bled white by the hecatomb of 1914–1918, and the French school became turned in on itself, grouped around its older members. Germany, on the other hand, succeeded better in preserving the lives of its scientists and in maintaining its great tradition of universality, assuring an exceptional degree of influence to its schools of mathematics. Students from many countries went there to be trained, notably young Frenchmen eager to renew ties with traditions forgotten in their own country since the death of Poincaré in 1912. After 1933, however, the flowering of the German and Italian schools was to be brutally arrested by fascism. Only after 1950 were they to be reestablished, influenced this time (by a curious reversal of the situation) by French mathematicians of the "Bourbaki" tendency. In Poland the mathematical schools were physically annihilated, since half the mathematicians were massacred by the Nazis. They did not recover their standing until after 1970. As for the English and Russian schools, they were able to come through the ordeal without suffering much damage.

Finally, it is appropriate to dwell on the establishment of schools of mathematics in the United States, for this illustrates the difficulty of implanting a tradition of research where one does not exist, even in a country rich in people and in resources. Up until 1870, no creative mathematician of any renown made his presence felt there; the development of a continent left scant room for abstract speculation. After 1880, the first efforts to create centers for mathematics consisted in inviting certain European (especially English) academics to come and teach in the universities, in founding periodicals devoted to mathematics and in sending gifted young students to be grounded in European mathematics. These efforts were crowned with success from about 1900. The first school to win international notice being that of Chicago, followed between 1915 and 1930 by those of Harvard and Princeton. An unexpected reinforcement came from the mass emigration of European mathematicians driven out by totalitarian regimes. It was they who were to contribute powerfully to the flowering of the very brilliant home-grown American schools of our day, which have placed themselves right in the fore by sensational discoveries, in group theory, algebraic topology and differential topology, among others.

The same slow development of schools of mathematics has taken place in India and in Latin America, but with more difficulty because of political and economic upheavals. China too, after 60 years of troubles, seems also to be engaged in this process, which could soon result in the arrival on the scene of some highly talented mathematicians, judging from those who were formerly able to pursue their careers outside their own country. However, many countries in the process of development have not yet succeeded in founding a lasting school. Further, in this as in all the sciences, it is necessary to reach a "critical mass", and except in exceptional cases (the Scandinavian countries since 1900, Hungary between 1900 and 1940) countries which are too small cannot hope to have a true national school, and need to attach themselves somehow or other to those of their more populous neighbors.

The attraction of large and active mathematical schools is easy to understand. The young mathematician on his own can quickly become discouraged by the vast, extent of a bibliography in which he is all at sea. In a major center, listening to his masters and his seniors, as well as to the visitors from abroad who throng in, the apprentice researcher will soon be in a position to distinguish what is essential from what is secondary in the ideas and results which will form the basis of his work. He will be guided toward key works, informed of the great problems of the day and of the methods by which they are attacked, warned against infertile areas, and at times inspired by unexpected connections between his own research and that of his colleagues.

Thanks to these special centers of research and to the network of communications which links them together over the whole planet, it is hardly likely that there could be the same lack of understanding in our day as certain innovators have had to bear in the past. In fact, as soon as an important result is announced, the proof is everywhere, to some extent, busily scrutinized and studied in the months which follow.

Jean Dieudonné was born in Lille in 1906. He received his doctorate from the École Normale Supérieure in 1931 and was a member of the faculty of universities in France, Brazil, and the United States. He was elected to the French Academy of Sciences in 1968. He died in 1992.

2
Why Is Mathematics?

A good question, to which there are at least two answers. One is that mathematics is a wonderful intellectual structure, endlessly fascinating and worthy of investigation for its own sake. Another is that of Morris Kline, the author of the following selection, who, though admitting the fascination of mathematics, thinks that it is first and foremost a subject whose purpose is to solve problems about the physical world. Other answers, such as the purpose of mathematics is to furnish employment for teachers, will not be considered.

The first two points of view provide alternative answers to the question "What is mathematics?" Many mathematicians, maybe most, are naive Platonists. That is, they think they know that the mathematical objects with which they deal were not made up by them, nor by anybody else. Mathematicians contemplate the objects, they wrestle with them and sometimes the objects fight back: clearly they have independent existence. It is easy to be a Platonist about the positive integers, which certainly seem to exist independent of whatever humanity chooses to think about them. More abstract structures have the same property. For example, mathematicians who have internalized Liouville's Theorem can *feel* the external existence of algebraic numbers. The theorem says that they're *different*—they stay away from rational numbers, refusing to cosy up to them the way that transcendental numbers do. They don't do that because our minds think they do, they do it because that's the way they *are*. Those italics in the preceding sentences express the difficulty of expressing the unexpressible, but it is the best that we naive Platonists can do. I myself have never felt the physical existence of, say, fibre bundles, but that is because I have never gotten close enough to them. Were I to become involved with them, I am sure that they would let me know that they are as real as the positive integers.

The other point of view is that mathematical objects are, obviously, creations of people. Where else could they possibly have come from? Mahler's S- and T-numbers clearly came from the head of Kurt Mahler, even as Mahler's Eighth Symphony came from the head (and heart) of Gustav Mahler, and when the theorem that T-numbers exist was proved, they existed only in the sense that an English word with five consecutive pairs of double letters exists. The word is "moonnookkeeper"—it denotes a person who tends to nooks on our satellite—and no one would argue that it is not the product of the human mind, or that moonnookkeepers exist in some world of Platonic ideals.

There is no doubt about what Morris Kline believed. The main job of mathematics, he said, is to solve problems in science. As we saw, Dieudonné would not agree. Here are two excerpts, from the beginning and the end of *Mathematics and the Physical World*.

Mathematics and the Physical World

Morris Kline (1959)

Preface

> In every department of physical science there is only so much science, properly so-called, as there is mathematics.   —Immanuel Kant

Mathematics is a model of exact reasoning, an absorbing challenge to the mind, an esthetic experience for creators and some students, a nightmarish experience to other students, and an outlet for the egotistic display of mental power. But historically, intellectually, and practically, mathematics is primarily man's finest creation for the investigation of nature. The major concepts, broad methods, and even specific theorems have been derived from the study of nature; and mathematics is valuable largely because of its contributions to the understanding and mastery of the physical world. These contributions are numerous.

Insofar as it is a study of space and quantity, mathematics directly supplies information about these aspects of the physical world. But, going far beyond this, mathematics enables the various sciences to draw the implications of their observational and experimental findings. It organizes broad classes of natural phenomena in coherent, deductive patterns. And today mathematics is the heart of our best scientific theories, Newtonian mechanics, the electromagnetic theory of Maxwell, the Einsteinian theory of relativity, and the quantum theory of Planck and his successors. Indeed, physical science has reached the curious state in which the firm bold essence of its best theories is entirely mathematical whereas the physical meanings are vague, incomplete, and in some instances even self-contradictory. Science has become a collection of mathematical theories adorned with a few physical facts. Further, if one can speak of *the* goal of modern scientific theory it is to subsume all its results under one mathematical principle whose implications would describe the multifarious operations of nature.

It is true that great mathematicians have often transcended the immediate problems of science. But because they were great and therefore understood fully the meaning of mathematics, they could discern directions of investigation which must ultimately prove significant in the scientific enterprise or shed light on the mathematical concepts that were already clearly instrumental in the investigation of nature. With this understanding and assurance they felt warranted in pursuing their own ideas and occasionally left the applications to others.

Unfortunately the relationship of mathematics to the study of nature is not presented in our dry and technique-soaked textbooks. Moreover, the fact that mathematics is valuable primarily because of its contributions to the understanding and mastery of nature has been lost sight of by some present-day mathematicians who wish to isolate their subject and offer only an eclectic study. An undue emphasis on abstraction, generality, rigor, and logically perfect deductive structures has caused a number of mathematicians to overlook the real importance of the subject. In the last fifty years a schism has developed between those who would hew to the ancient and honorable motivations for mathematical activity—the motivations which have thus far supplied the substance and most fruitful themes—and those who, sailing with the wind, would investigate what strikes their fancy. But history favors only one view. To supply the historical evidence within the limits of a preface it must suffice to mention that the greatest mathematicians —Eudoxus, Apollonius, Archimedes, Fermat, Descartes, Newton, Leibniz, Euler, Lagrange, Laplace, Hamilton, Cauchy, Gauss, Riemann, and Poincaré, to mention just a few—were also giants of science. All of these men would have earned enduring fame for their physical researches alone.

To display the role of mathematics in the study of nature is the purpose of this book. Subordinate, but by no means incidental, objectives may also be fulfilled. We may see mathematics in the process of being born—that is, see how physical problems, when idealized and formulated in the language of number and geometry, become mathematical problems and see how, guided still by intuition and physical thinking, the mathematician creates a new method or a new branch. The precise manner in which mathematics produces answers to physical problems —mathematics at work, so to speak—may also be evident. As we follow the gradual development of mathematical power and the increasing absorption of mathematics in the scientific enterprise, we may learn how and why mathematics has become the essence of scientific theories. Though mathematics has been intimately involved with science—on a Platonic basis of course—mathematics is not physics, astronomy, or chemistry. Hence the distinctions between mathematics and science will also be presented.

Finally, a study of mathematics and its contributions to the sciences exposes a deep question. Mathematics is man-made. The concepts, the broad ideas, the logical standards and methods of reasoning, and the ideals which have been steadfastly pursued for over two thousand years were fashioned by human beings. Yet with this product of his fallible mind man has surveyed spaces too vast for his imagination to encompass; he has predicted and shown how to control radio waves which none of our senses can perceive; and he has discovered particles too small to be seen with the most powerful microscope. Cold symbols and formulas completely at the disposition of man have enabled him to secure a portentous grip on the universe. Some explication of this marvelous power is called for. A presentation of these features of mathematics may also make clear the true meaning of the subject and the reason that it is regarded as of supreme importance in human thought.

Mathematics and Nature

How can it be that mathematics, a product of human thought independent of experience, is so admirably adapted to the objects of reality?

—Albert Einstein

There is a well-known story, largely apocryphal, concerning the visit of the French Encyclopedist, Denis Diderot, to the court of Catherine the Great. During his stay there he shocked many members by his atheistic views. To silence him the Empress decided to play a royal trick. She informed him that a great mathematician, Leonhard Euler, would present to the court a mathematical demonstration of the existence of God and invited Diderot to be present. Diderot accepted. Euler appeared, rattled off some meaningless mathematical formulas, and concluded, "So God exists." He then contemptuously challenged Diderot to refute him. According to the story, which is incorrect in this detail, Diderot knew no mathematics and was so disconcerted that he withdrew shamefaced from the court. Presumably the joke was on Diderot; actually the joke was on Euler and the entire mathematical world.

Every mathematician of the eighteenth century was sure that his subject proffered the most certain body of truths man possessed. Euler was perhaps the cockiest of this group. At the age of twenty, with no practical experience to guide him, he wrote a theoretical paper on the design of ship masts, which, incidentally, won a prize offered by the Paris Academy. He stated in his conclusion that he did not think it necessary to check his results by experiments because they had been deduced from the surest foundations in mathematics and mechanics, hence their correctness could not be questioned. The conviction was so widespread that mathematics contained within itself the profound inner design of the universe that many mathematicians, including Euler, believed that the existence of God could be proved as a necessary consequence. Indeed, Euler spent many months in trying to establish this conclusion by diverse mathematical arguments.

But, as Francis Bacon had already observed, "Nature is more subtle many times over than the senses and understanding of man." The development of non-Euclidean geometry showed that man's mathematics did not speak for nature, much less lead to a proof of the existence of God. It became clearer, too, that man institutes the order in nature, the apparent simplicity, and the mathematical pattern. Nature itself might have no inherent design, and perhaps the best one could say of man's mathematics was that it offered no more than a limited, workable, rational plan.

But if mathematics lost its place in the citadel of truth, it still was at home in the physical world. The major, inescapable fact, and the one that still has inestimable importance, is that mathematics is the method par excellence by which to investigate, discover, and represent physical phenomena. In some branches of physics it is, as we have seen, the essence of our knowledge of the physical world. If mathematical structures are not in themselves the reality of the physical world, they are the only key we possess to that reality. The discovery of non-Euclidean geometry not only did not destroy this value of mathematics or confidence in its results but rather, paradoxically, increased its usefulness, because mathematicians felt freer to investigate radically new ideas, and yet found these to be applicable. In fact the role of mathematics in the organization and mastery of nature has expanded at an almost incredible rate since 1830. And the accuracy with which mathematics can represent and predict natural occurrences has increased remarkably since Newton's day. We seem, therefore, to be confronted with a paradox. A subject that no longer lays claim to truth has furnished the marvelously adaptable Euclidean geometry, the pattern of the extraordinarily accurate heliocentric theory of Copernicus and Kepler, the grand and embracing mechanics of Galileo, Newton, Lagrange, and Laplace, the physically inexplicable but marvelously applicable electromagnetic theory of Maxwell, and the sophisticated theory of relativity of Einstein. All of these highly successful developments rest on mathematical ideas and mathematical reasoning. The question becomes inescapable. Why does mathematics work?

From the time of the Greeks, when the power of mathematics to reveal new knowledge of the physical world was first glimpsed, this question was considered and answered according to the light of the age. Let us examine some of these answers. They are not only of interest but are still the explanations offered by some people. Insofar as the axioms of mathematics are concerned the dominant view until recent times was that they were self-evident truths about the physical world. That there were truths in general was unquestionable, and the truths of mathematics were so apparent that one needed no experience to recognize them. Plato went further than most philosophers by asserting that the true reality consisted of ideas and relationships among them; the physical world was merely an imperfect representation of these ideas and relationships. Among philosophers who accepted truths, Descartes was most critical. But he was confident of the mathematical truths and accounted for his certainty with the argument that God planted within us the ideas of numbers and figures, and from the knowledge of these ideas their basic properties were apparent. Though Galileo certainly tried to rid himself of preconvinced notions he, too, was sure that there are truths and that we can discover them by paying attention to what nature says. Of such truths the axioms of mathematics were the ones most readily grasped.

The greatest eighteenth-century philosopher, Immanuel Kant, was as certain as these other men of the truths of mathematics and of the axioms in particular; however, Kant rested his case on the thesis that the human mind supplies the concepts and axioms with which to organize experience. Hence there was a necessary precise correspondence between what the mind accepts and experience demonstrates. Kant did make a distinction between the knowledge that the mind builds up with the sensations it receives and the physical world that, so to speak, lies beyond the sensations and is unknowable, but he was confident that the mind's organization of knowledge gave us reliable and indeed infallible knowledge.

Though the axioms of mathematics were accepted self-evident truths, there remained the larger question of why the deductive arguments performed by human reason yielded truths. In other words, why should there be agreement between the theorems and physical experience? Might not the reasoning lead away from truths? Prior to the advent of non-Euclidean geometry the almost universal answer to this question was that the universe was rationally and indeed mathematically designed. Of course the designer was God. Had Plato, for example, written the Bible it would undoubtedly have opened with the statement: "In the beginning God created mathematics and then created heaven and earth according to the laws of mathematics." Copernicus and Kepler saw the hand of God in the marvelous correspondence between their mathematical scheme and the actions of nature. Galileo expressed his belief in the mathematical design of nature in a classic statement: "Philosophy is written in that great book which ever lies before our eyes—I mean the universe—but we cannot understand it if we do not first learn the language and grasp the symbols, in which it is written. This book is written in mathematical language, and the letters are triangles, circles, and other geometrical figures, without which it is humanly impossible to comprehend a single word." Newton saw everywhere in the universe evidence of God's majestic design and of His constant and continuous concern to keep the universe running according to plan. As James Thomson, in his memorial poem on Newton, put it, Newton,

> *... from motion's simple laws*
> *Could trace the secret hand of Providence,*
> *Wide-working through this universal frame.*

Leibniz accounted for the agreement between mathematical reasoning and facts by "the pre-established harmony between thought and reality." Nature was the art of God.

Of course the mathematical design of nature had to be uncovered by man's continual search. God's ways appeared to be mysterious but it was certain that they were mathematical, and that man's reasoning would in time discern more and more of the rational pattern which God had utilized in creating the universe. The fact that man reasoned exactly in the way in which God had planned seemed readily understandable on the ground that there could be but one brand of correct reasoning. The possibility that man's reasoning might not conform to God's ratiocinations occurred to Descartes, but he dismissed it with the argument that God would not deceive us by allowing our mental faculty to function falsely.

Despite the persistence of the belief in a rational universe designed by God, some contrary views cropped up even before the discovery of non-Euclidean geometry. The intensely religious and yet occasionally wavering Blaise Pascal remarked that nature proves God only to those who already believe in Him. When Napoleon, on being handed a copy of Laplace's *Celestial Mechanics*, remarked to the author that there was no mention of God in this great and extensive work, Laplace replied, "I have no need of this hypothesis." However, Laplace had not considered the question of how it was that his superb mathematical reasoning should fit the universe so remarkably well.

The creation of non-Euclidean geometry thrust to the fore the question of how it is that mathematical reasoning gives us knowledge of the physical world. Since reasoning on the basis of axioms contradicting Euclid's yielded theorems that applied to the physical world, it was no longer possible to assert that mathematics yielded truths, for truth is unique. Nevertheless, because the applicability of mathematics to the physical world remained as effective as before, and new developments added to the extent and depth of these applications, some great mathematicians and scientists, notably Sir James Jeans and Sir Arthur Stanley Eddington, continued to affirm God's mathematical design of nature. About the only difference in this later view from those of the seventeenth and eighteenth century is that one can no longer affirm that the present mathematical account of the heavenly motions, for example, or of electromagnetic waves, is the final one but rather an excellent approximation which, by continued improvement, may become the correct formulation.

It seems fair to say, however, that within the last century the belief in the mathematical design of nature has receded. Yet the problem of accounting for the remarkable insight into the behavior of nature that mathematics offers still remains. Peculiarly enough, the effect of non-Euclidean geometry on our aggravated problem makes it easier to approach an answer, even if the resulting explanation may not be wholly satisfying. What we have to explain now is not how mathematics produces truths but rather the correspondence between physical reality and the mathematical representation.

We shall attempt to answer this question in the light of our twentieth-century understanding of the nature and role of mathematics. Mathematics begins with the selection of certain concepts that appear to be instrumental in studying the physical world, for example, the concepts of number and geometry. Having chosen his concepts the mathematician seeks next some sound fundamental facts on which to base his reasoning. There is not much doubt that these premises or axioms are inferred from observations and experience. Actual experience with collections of physical objects shows that $3(4+5) = 3 \times 4 + 3 \times 5$. Consequently, the mathematician adopts as an axiom about number the general statement, called the distributive axiom, which says that for any numbers a, b, and c, $a(b+c) = ab + ac$. To men whose experience with physical space was limited to small regions it seemed clear that a straight line could be extended indefinitely in either direction. This experience was so common that it did not occur to man to question, before the development of non-Euclidean geometry, whether such an axiom applies to the entire three-dimensional universe.

In the most fruitful applications of mathematics to the physical world, some nonmathematical axioms also enter. The Newtonian system of mathematical mechanics depends as much on the Newtonian laws of motion and gravitation as it does on the axioms of mathematics. The experiential basis for the physical axioms is more evident. Yet one cannot really distinguish the physical axioms from the mathematical ones on logical grounds. Both types are suggested by observation and experience and are abstractions from experience. It is true that the axioms of number and geometry were obtained first and have existed, therefore, for many more centuries. They may even be in some vague sense more fundamental, that is, the quantitative and geometrical properties may be the basic ones, and the others of secondary importance, just as food is a primary need, whereas shelter, at least in some climates, is secondary. If anything, we might say that the mathematical axioms are more obvious, whereas the correct physical axioms require a deeper and more penetrating analysis of the physical world to be uncovered. But this distinction is unimportant because both types are needed to make any real progress in the study of the physical world. Nevertheless, we shall stress the role of the mathematical axioms because our problem at the moment is to account for the effectiveness of mathematics.

The third element in the mathematical approach to nature is the obtainment of new conclusions. In this phase of his work the mathematician may indeed be distinguished from the physical scientist. The mathematician, as we have already noted many times, insists on establishing his conclusions by deductive arguments, whereas the scientist will feel free to resort to observations and experimentation and subsequent inductive generalizations to arrive at his conclusions. With concepts and axioms in mind, whether solely mathematical or a combination of physical and mathematical ones, the mathematician retires to a corner and deduces new conclusions about the physical world. These conclusions generally depend upon hundreds of steps of pure reasoning and yet yield such knowledge as the distance to the sun and sometimes such totally unexpected phenomena as the existence of radio waves. It may indeed be true that the deduced facts are necessary consequences of the axioms, and the latter derive from the physical world, but that reasoning of a highly intricate type should produce physically serviceable knowledge is the mystery which demands resolution. Why should the physical world conform to the pattern of man's reasoning?

One widely accepted view is that man learns to reason by studying nature.

Let us see just what this means. One of the basic laws of reasoning says that an assertion cannot be both true and false. One cannot assert that the sun is both red and non-red. In the mathematical realm one says, for example, that two lines cannot be both parallel and nonparallel. This law is known in logic as the law of contradiction. What prompts us to accept this principle of reasoning? The answer is that this is what we observe in nature. Objects do not possess contradictory qualities at the same time. Of course the reader may protest, "How could it be otherwise? Is it not nonsense to believe that an object can be both red and non-red?" It is true that we cannot conceive of such an object, but this inability may be due to the superficiality of our thinking or to our training. The mathematicians and scientists who identified Euclidean geometry and space could not conceive of any other geometry. Their argument was that there is but one space and one body of laws about that space and the two agree perfectly. Many educated people say today that $2 + 2$ must be 4. But there are algebras, physically useful algebras, in which this statement does not hold.

If at the present time at least the law of contradiction appears to be inescapable, rather than an inference from experience, there are other logical principles that appear more clearly to have an empirical origin. It is correct to argue that if all angels are white and no salamander is white, then no salamander is an angel. This argument is deductive and valid. Nevertheless, until one acquires experience with the logical principle used in this argument, and in fact even tests the conclusions of arguments made with the principle, he is not usually inclined to accept it. The doubt is understandable. Let us note that the seemingly similar argument, if all angels are white and no salamander is an angel, then no salamander is white, is not valid. Hence it seems very likely that man has learned to reason by studying what happens in nature, and it is therefore not surprising that his reasoning yields results that accord with nature.

History affords some confirmation of this thesis. The laws of reasoning were formulated abstractly by Aristotle in the fourth century BC, centuries after man had practiced precise reasoning. Moreover, the mathematical reasoning of the preceding centuries concerned readily visualizable geometrical facts so that observation and measurement could readily have been employed to correct or improve poor deductive reasoning.

Whether or not man's reasoning is consistent with the behavior of nature, he has one more means at his disposal to make his mathematics fit the physical world. If his theorems do not fit, he is free to change his axioms. This recourse seems farfetched, but scientists have adopted this procedure in our time. The increased range of astronomical observations and the study of motions with speeds approaching that of light caused Einstein to adopt a non-Euclidean geometry for his theory of relativity. The mathematician may have to change his axioms still further as he attempts to apply them to more diverse or to more accurately measured phenomena.

There are other philosophical explanations of the remarkable correspondence between mathematical reasoning and the behavior of the physical world, but it must be admitted that no one of these is final. It may be that the effectiveness of the mathematical representation and analysis of the physical world is as unexplainable as the very existence of the world itself and of man. It is nonetheless a phenomenon that we can accept and utilize to great advantage. Having found that this gift from an unknown donor can be so profitably employed, our civilization now seems set to exploit it to the hilt. If the attempt to understand why it works leaves us with an enigma, as does the smile of the Mona Lisa, this merely means that we have an intriguing subject for further study and contemplation.

The dominant view today as to the nature of mathematical activity is, then, that the concepts and axioms are derived from experience, the principles of reasoning used to deduce new conclusions were most likely derived from experience, and, insofar as the applicability of mathematics to the physical world is concerned, the conclusions must be checked against experience. Except for the fact that mathematics insists on deductive proof of its conclusions, it would appear that this subject hardly differs from any branch of physical science. Shall we conclude, then, that mathematics is just a special kind of science?

The answer is no. The description we have just given of the subject and the aspects we have emphasized throughout this book afford too narrow a view. Even the Greek mathematicians, who certainly believed that they were pursuing truths of nature, felt free to investigate ideas that had no immediate application but nevertheless fitted their conception of legitimate mathematical activity. A classic example is afforded by the conic sections. Fifteen hundred years before Kepler used the ellipse in astronomy and Galileo used parabolas to describe trajectories, the Greeks had explored these curves fully.

The history of non-Euclidean geometry provides another example of the exploration of an idea that was not intended for use in science or engineering. The hundreds of investigators who sought to replace Euclid's parallel axiom by an intuitively more acceptable one were not trying to correct any errors in Euclidean geometry and certainly did not even conceive of a new geometry. Their concern was largely to perfect the structure of geometry. And even though Gauss, Bolyai, and Lobatchevsky had suspected that Euclidean geometry was not necessarily the geometry of physical space, and even though they mentioned the possibility of their new geometry being applicable to astronomy, they explored this radical idea more for the sake of what light it might shed on Euclidean geometry. The history of such creations shows that while nature is the womb from which mathematical ideas are born, these ideas can be studied for themselves while nature is left behind. Hence one must recognize that mathematicians have always felt free to pursue ideas that appeared to them to be relevant to their subject, regardless of whether the resulting work would be of immediate value to science.

It is true of the work we have just discussed that the mathematicians undoubtedly thought that it at the very least rounded out knowledge of subjects important for the study of nature. Nevertheless, this work shows that one must include in the range of mathematical investigations ideas that are only indirectly related to ones which prove useful. The freedom to explore such ideas has proved to be a boon in several respects. The very ideas we mentioned above did ultimately yield a new and richer insight into nature. Mathematicians have thereby not only anticipated the needs of science but have suggested the directions science should take. It seems unlikely that Kepler would have invented the ellipse to describe planetary motion, because that task, together with the one he actually performed of fitting the ellipse to data, would have been superhuman. Most likely he would have modified the theory of epicycles still further. Likewise, Einstein took full advantage of the already existing non-Euclidean geometry to create the theory of relativity. Undoubtedly the tasks of conceiving of such a geometry, which in itself demanded the genius of Gauss, and applying it to a radically new physical theory would have been beyond the powers of one man.

But it would be wrong to suppose that mathematicians have felt obliged to limit themselves to ideas possessing potential or indirect bearing on science. The individual can be motivated by forces that have little to do with the broader social and historical movements directing the larger currents of thought. Many mathematicians pursue the subject simply because they like it. To them the subject offers intellectual challenge and values that draw them to it far more strongly than money or power attract people generally. They enjoy the excitement of the quest for new results, the thrill of discovery, the satisfaction of mastering difficulties, and the pride in achievement. There are, moreover, delights and aesthetic values to be derived from surveying orderly chains of reasoning such as occur in most proofs, from the contemplation of the results themselves, and from grasping the ideas that make the proofs work. Those portions of mathematics which prove valuable in the study of nature offer the additional satisfaction of unifying a multitude of seemingly disorganized facts and of comprehending nature's ways.

One finds evidence of the aesthetic drive and the appreciation of mathematical beauty in the writings of many mathematicians. Archimedes, for example, though famous for his scientific work and inventions as well as for his original and powerful mathematical work, valued the latter more. He did not deign to leave any written account of his inventions. Isaac Newton spoke of God as interested in the preservation of cosmic harmony and beauty and regarded the mathematical design of the universe as an expression of that beauty.

The mathematician is free to create beautiful theorems and theories. But one might well ask whether there is any test of this beauty. Unfortunately there are no truly objective criteria, any more than there are in art or literature. It is fair to require that a valuable mathematical creation use effectively the axioms on which it rests; that is, one must not use a blunderbuss to kill a flea. The reasoning should be sufficiently powerful or complex; a trivial or immediately obvious argument hardly warrants being written down. The most desirable proofs contain some underlying scheme or display some principle or method that is applicable to other problems than the one in question. And there should be in addition some idea or device that produces surprise or gratification that a genuine difficulty has been overcome. But all of these criteria are basically subjective, and a piece of work that the creator or some readers may regard as superb can be dismissed contemptuously by others. Since mathematicians are human beings there is no doubt that jealousies, rivalries, and even closed-mindedness have caused some to deprecate fine creations. A standard joke, "Trivial is what the other fellow does," jibes at the all too common failure of mathematicians to be sufficiently appreciative of the work of others.

To recapitulate the substance of our last few remarks, we may say that the mathematician creates theory that is designed to solve current problems of science, theory that sheds light on the physically applicable results though not directly of use, theory that has potential use in science, and theory that offers aesthetic values. We see, therefore, that the scope of the mathematician's work is considerably broader than what has been presented in this book.

Mathematicians are free to pursue their own inclinations and to cater to their own tastes, but there is, of course, the danger that ideas pursued solely for their intrinsic appeal may deviate so far from normal human interests that they will attract only an esoteric group. There is also the danger that ideas concocted solely by the human mind and bearing no relation to the meaningful and weighty phenomena of the physical world may become thin, impoverished, and insignificant however facile the reasoning which develops the ideas. In view of the vast extent of mathematical knowledge and the proliferating demands of science, there is much already done and yet to be done that offers all the satisfactions of arbitrarily chosen directions of research and yet has more substance and more moment for our civilization. Certainly the lesson history teaches, and the reason for the great emphasis placed on mathematics today, is that mathematics provides the supreme plan for the understanding and mastery of nature.

Mathematics may be the queen of the sciences and therefore entitled to royal prerogatives, but the queen who loses touch with her subjects may lose support and even be deprived of her realm. Mathematicians may like to rise into the clouds of abstract thought, but they should, and indeed they must, return to earth for nourishing food or else die of mental starvation. They are on safer and saner ground when they stay close to nature. As Wordsworth put it, "Wisdom oft is nearer when we stoop than when we soar."

Morris Kline (1908–1992) spent his teaching career at New York University. He was the author of several books, including *Mathematics in Western Culture* (1953), *Mathematical Thought from Ancient to Modern Times* (1972), and *Mathematics, the Loss of Certainty* (1982).

He was a forthright non-Platonist, saying

> Mathematics does appear to be the product of human, fallible minds rather than the everlasting substance of a world independent of man. It is not a structure of steel resting on the objective bedrock of objective reality, but gossamer floating with other speculations in the partially explored regions of the human mind.

He also wrote *Why Johnny Can't Add* (1973), a criticism of the New Math movement of the 1960s and its emphasis on structure and the abstract. He was correct that the New Math did not improve the state of mathematics education in this country, but he was perhaps not so correct about what he thought would. In an address to the Mathematical Association of America in 1955, when the New Math was on the horizon, he said

> Let us look at the status of mathematical education. I believe that I do not have to convince anyone here that we are failing to put mathematics across. One only has to note the reaction of students to the subject, for example, their grim countenances in class, to see that we are failing. One can check this conclusion by asking his colleagues in other departments—surely an intelligent group—how they feel about the mathematics they took at school.

Fifty years later, exactly the same words could be said: we must still be failing. (In my view, we are not failing now, nor were we fifty years ago. The fact that nothing seems to have changed over that length of time is evidence for that.) Kline's solution to the problem was to change the curriculum and make it more meaningful by concentrating on solving problems in the physical world. I think that countenances would still be grim—perhaps not the same ones, but just as many.

3
Is Mathematics Inevitable?

How important are we? The answer to that is easy: with probability very close to 1, none of us has any importance at all in the world-historical sense. Even locally, we may not matter as much as we think we do. As G. H. Hardy said in his *Mathematician's Apology*, "But solicitors and stockbrokers and bookmakers often lead comfortable and happy lives, and it is very difficult to see how the world is richer for their existence." But what about those who are not as ordinary as we? What about Gauss, Newton, Archimedes? Did they make a difference?

They did, of course, or else we would not remember them. But, were they necessary? Gauss had to make a choice between mathematics and philology and chose mathematics; if he had gone the other way, would we have the quadratic reciprocity theorem and congruence notation? There is no way of knowing, but in the following selection, Nathan Altshiller Court argues that the answer is "yes". If our great ones had not existed, other great ones would have done what they did.

I am not so sure. This sounds like the view that we are the inevitable climax of history, and all the laborings of past generations were for the purpose of producing the world in which we live. This is quite egotistical and shows a lack of imagination—other arrangements are possible. If the British code-breakers had failed in the second world war, Germany could have won and the world would be quite different. John Wilkes Booth might have missed. The apple might not have landed on Newton's head.

It is probably the case that calculus would still have appeared even if Newton had died at birth and Leibniz had never thought about it. There was proto-calculus in the work of Fermat, Wallis, and others, and someone would have carried on from where they left off. However, no one can know for sure. If Benoit Mandelbrot hadn't thought of fractals, would anyone else have? My instinct tells me that it is possible that the world could have remained forever fractalless.

Questions like this can never be answered. Speculation about the answers does no harm, though, and can do some good by stimulating us to think about some things that are worth thinking about. See what you think of the following.

Mathematics in Fun and in Earnest

Nathan Altshiller Court (1958)

Mathematics and Genius

A. The "Heroic" and the "Objective" Interpretations of History. Once upon a time, so the story goes, a beautiful stallion was brought to the royal court and presented to the king. The stallion was very wild. The king was warned that no man had ever managed to mount the fiery beast. The heir apparent who happened to witness the presentation ceremony of this unusual gift, jumped upon the back of the spirited horse, and before anybody had time to realize what was happening, the young prince was already way out of sight. The King's anxiety for the safety of his beloved son was very great. After a certain lapse of time, the young man reappeared, safe and sane, on the back of the subdued, tame animal. The proud and loving father was so elated that he exclaimed in exaltation: "My son, find for yourself another kingdom. Mine is too small for you." These accidental words of the king took deep root in the sensitive soul of the young prince. History knows this young man under the name of Alexander the Great (356–323 BC), the famous conqueror of the ancient world.

I read this story in my school-text on ancient history, a fine book, full of names and dates. Every historical event had its precise moment of occurrence recorded. You were told exactly by what king, or general, or by what great leader any given event was brought about. For the sake of brevity let us refer to this way of conceiving historical events as the "heroic view" of history.

This heroic interpretation of history is very attractive, because of its simplicity and its definiteness. All the whys and wherefores are readily answered by the names of the great men who made the history of the nation, or of the race. However, this heroic view has an obvious weakness: it makes history whimsical, capricious, and accidental, to the point of triviality. Suppose that our stallion of a moment ago, in its frantic effort to rid itself of its unsuccessful and unlucky tamers, had broken a leg, or two. King Phillip would have been deprived of the occasion to utter those fateful words of his, and his son Alexander would have lived out his life as an obscure and inconsequential ruler of the little kingdom of Macedonia.

According to a much repeated saying, of undetermined origin, "God made George Washington childless, so he could become the father of his country." Thus, if it were not for some physiological peculiarity or deficiency of Martha Washington (or was it of George himself?) this country would have remained a British colony, even unto this very day and generation.

During the nineteenth century various writers, like the Englishman Henry Thomas Buckle (1821–1862), the Frenchman Hippolyte Taine (1828–1893), best known in the English-speaking world for his history of English literature, and the German Karl Marx (1818–1883) have advanced the view that human history is not made by individuals, but is dominated by objective factors, like climate, geographic environment, natural resources, economic and social conditions, etc. This objective interpretation of history has since gained a great deal of ground. A forceful presentation of this conception may be found in the presidential address delivered before the American Historical Association by Edward P. Cheyney (1861–1947), under the title "Law in history" [1] in which the following two passages occur: "History, the great course of human affairs, has not been the result of voluntary action on the part of individuals or groups of individuals, much less of chance, but has been subject to Law." "Men have on the whole played the part assigned to them: they have not written the play. Powerful rulers and gifted leaders have seemed to choose their policies and carry them out, but their choice and success with which they have been able to impose their will upon their time have alike depended on condition over which they have had no control".

The heroic and the "objective" interpretation of history are obviously poles apart. Which of them is right? General human history is so many-sided, so complex, that it is easy enough to emphasize one element or another of its vast contents and arrive at conclusions which are contradictory, and still have each a good deal of truth in them. We may try to simplify the problem, as we often do in mathematics, reduce the number of variables, and examine a few of them at a time.

Our objective may perhaps be achieved more readily if we examine the history of a restricted, particular domain, say, that of mathematics.

B. Are Inventions Inevitable? We are accustomed to pronounce with respect and admiration, not to say with reverence and awe, names like Euclid, Archimedes, Descartes, Newton, Leibniz, Lagrange, Gauss, Poncelet, Klein, Poincaré, and many others. We know the books those men have written, the theorems which bear their names. In our own time we know by name men who live in our midst and some of whom we know personally, men who lend luster and glory to our generation, men who give us courage and inspiration. Through the study, direct and indirect, of the works of these eminent scholars we know what they have contributed to the growth and advancement of mathematical science. There hardly can be a more forceful confirmation of the importance of the individual in history, of the heroic interpretation of history, if you will. Nevertheless, there is another side to this medal.

On December 21, 1797, in Paris, the great mathematicians Laplace and Lagrange were both present at a brilliant social gathering which included a great many celebrities. Among the guests was also a victorious young general whose star was ascending rapidly, and who happened to be a former student of Laplace. In the course of the evening the general, while talking to the two world famed scholars, entertained them with some unusual and curious solutions of well known problems of elementary geometry, but solutions with which neither of his two eminent listeners were familiar. Laplace, a bit peeved, finally said to his erstwhile pupil, "General, we expect everything of you, except lessons in geometry". The name of the young general was Napoleon Bonaparte. Napoleon had learned about those strange constructions during his famous campaigns it Italy, whence he had just returned. While there, he met Lorenzo Mascheroni, a professor at the University of Pavia, who that very year, 1797, published a book *Geometria del Compasso* in which the author showed that all the constructions that can be carried out with ruler and compass, can also be carried out with compass alone, a very astonishing result, indeed. Had Mascheroni died in infancy, would science have been deprived forever of those Mascheronian constructions? One may think the question preposterous, for such a hypothetical query admits of no answer, one way or the other. Curiously enough, in the present case the question can be answered, in a very definite way. A century and a quarter before the publication of Mascheroni's book a Danish mathematician Georg Mohr published in Amsterdam a book in two languages, one in Danish and the other in Dutch, simultaneously, in which he gives Mascheroni's main result, as well as the solutions of a good many of the problems solved later by the Italian scholar. Mohr's book passed entirely unnoticed by his contemporaries. It came to light in the present century by accident. In the preface to his book Mascheroni states explicitly that he knows of no previous work along the same lines as his book, and there is not the slightest reason to doubt his word.

The story emphasizes the fact that so many mathematical discoveries, great and small, have been made independently by more than one scholar. This multiplicity of claims to the discovery of one and the same thing is probably the most outstanding fact in the history of mathematics.

The dispute as to whether Newton or Leibniz invented the calculus is well known. The French claim, with a good deal of justice, that Fermat anticipated both of them. It is only Fermat's strange and persistent aversion to the pen that deprived him of the credit as inventor of that powerful mathematical tool.

A similar story may be related about the epoch-making discovery of analytic geometry. There is as much reason to refer to this discovery as "Fermatian" as there is to call it "Cartesian." Carl B. Boyer in the preface to his *History of Analytic Geometry* [2] says: "Had Descartes not lived, mathematical history probably would have been much the same, by virtue of Fermat's simultaneous discovery (of analytic geometry)."

The geometric interpretation of complex numbers was discovered independently and almost simultaneously by four different men, at the beginning of the nineteenth century. An instrument for drawing a straight line without the use of a ruler, known as the "cell of Peaucellier" (1832–1913), was also invented by a young student Lipkin of St. Petersburg (Leningrad) [3].

Even whole theories have grown up, the paternity of which nobody can claim with justice. A good and simple example of this kind is offered by the theory of inversion. This theory came into being early in the nineteenth century, and from so many different quarters that it is impossible to associate any particular name with it. The only thing that can be said about it is that, like Topsy, it "just growed."

The multiplicity of claims to the same discovery is so common that not only have we stopped to be surprised by it, but we have grown accustomed to expect it. Better to be able to protect the priority right of contributors, most of the editors of mathematical journals add to each article they publish, the date when that paper was received in the editorial office.

What was said about mathematics may be repeated with equal force about astronomy, physics, chemistry, mechanics, in fact about any science, pure or applied. Two industrious sociologists compiled a list of inventions, each of which has more than one claimant to its paternity. The list contains 148 entries and is far from being exhaustive. Armed with their incredible, but correct list, the two authors fire, point blank, an amazing question at their readers, namely: "Are inventions inevitable?" [4].

C. Genius and Environment. We are prone to think that the essence of genius is freedom. Does not genius invent or create what he will? On closer examination, however, it is seen that this conception of genius is an exaggeration. What a genius may accomplish depends upon circumstances which can be controlled by no individual. The invention of the creative individual is necessarily an extension of the knowledge of his time, or is something that satisfies the needs of his contemporaries. These characteristics have to be incorporated in the invention, if the genius is to be recognized as such. If a self-taught scholar from somewhere in the hinterland would send to the editor of a journal or to the Academy of Science a manuscript which in substance would amount to the discovery, say, of non-Euclidean geometry, or of the sextant, not much fuss would be made about the author, even if his honesty would not be called into question. And such things happen, on various levels of achievement. About the middle of the nineteenth century the Academy of St. Petersburg was offered by a teacher in some rural elementary school a crude exposition of the basic ideas of the calculus.

On the other hand, what genius can accomplish depends upon what others have done before. Newton realized that if he had seen farther than others, it is because he was "standing on the shoulders of giants." "Perhaps nowhere does one find a better example of the value of historical knowledge for mathematicians than in the case of Fermat, for it is safe to say that, had he not been intimately acquainted with the geometry of Appollonius and Viète, he would not have invented analytic geometry." [5]. On the other hand, as great a genius as Archimedes could not invent analytic geometry, for the algebraic knowledge necessary for such an achievement was not available in his time.

The relation between the genius and the culture he is born into is expressed by A. L. Kroeber in the following way [6]: "Knowing the civilization of a land and of an age, we can then substantially affirm that its distinctive discoveries, in this or that field of activity, were not directly contingent upon the personality of the actual inventors that graced that period, but would have been made without them; and that, conversely, had the great illuminating minds of other centuries been born in the civilization referred to instead of their own, its first achievements would have fallen to their lot. Ericsson or Galvani, eight thousand years ago, would have polished or bored the first stone; and in turn, the hand and mind whose operations set in inception the neolithic age of `human culture, if held in its infancy in unchanged catalepsy from that time until today, would now be devising wireless telephones and nitrogen extracts," or (let us add) nuclear weapons and interstellar ships, a generation or two later.

The dependence of the individual, whatever his natural endowments, upon the time and civilization he happens to live in, becomes quite obvious, once attention is called to this phenomenon. We are not a bit surprised to see that the French children are so very partial to the French language, and that the Chinese children, not to be outdone by the French, speak as unanimously the Chinese tongue. The same may be said, in a broader sense, about arts and crafts, music, or any other component element of culture. On a larger scale, analogous remarks may be made about those parts of culture which have become common to a considerable part of mankind, like the sciences, and mathematics in particular.

These observations may help us to comprehend the reasons for the multiplicity of claims for the same discovery. The anthropologist Leslie A. White puts it this way: [7] "In the body of mathematical culture there is action and reaction among the various elements. Concept reacts upon concept: ideas mix, fuse, form new syntheses. When this process of interaction and development reaches a certain point, new syntheses are formed of themselves. These are, to be sure, real events and have their location in time and space. The places are, of course, the brains of men. Since the cultural process has been going on rather uniformly over a wide area of population, the new synthesis takes place in a number of brains at once."

Tobias Dantzig (1884–1956) in his admirable book, *Number—The Language of Science* [8], says the same thing, with a different emphasis: "It seems that the accumulated experience of the race at times reaches a stage when an outlet is imperative and it is merely a matter of chance whether it will fall to the lot of a single man, two men, or a throng of men to gather the rich harvest."

D. Genius and the "Instinct of Workmanship". Granting that objective conditions determine the kind of discoveries that can be made at any given period of history on the one hand, and that on the other hand such inventions are "inevitable," such forward steps do not take place automatically. Each particular advance requires an effort, and often a very strenuous one, on the part of the gifted individual, the "genius" who brings it about. What impulse does the individual respond to, when he makes the requisite effort?

Mathematics, like any other science, in its early stages developed empirically for practical, utilitarian purposes. The demand for its services never cease through the ages, although the extent and the pressure may vary widely from one period to another, and mathematical inventiveness may vary accordingly. This is quite clear, for instance, in the case of the rapid strides made by mathematics during the brilliant seventeenth century. The mathematician, like any other scientist, is not unmindful of the needs of his time and is not indifferent to the acclaim that would be his if he supplied the answer to a pressing question of his day.

There is, however, another phase of the situation to be considered. After a sufficient amount of mathematical knowledge has been accumulated, the cultivation of this domain of learning may become an interest in itself. Those versed in its secrets and adept in manipulating them may find it attractive to strive for new results just to satisfy what Thorstein Veblen (1857–1929) called the "instinct of workmanship" [9]. The only extraneous element in the case may be the wish to gain the approval of the restricted audience of like-minded people or perhaps to confound some rivals. Outside of that the reward that may accrue to the mathematician for his efforts is to live through the pains of creation and to experience the exhilarating joy of discovery. His is a labor of love. He considers himself amply repaid if he feels that he added, be it ever so little, to the luster of the brightest jewel in the intellectual crown of mankind—The Science of Mathematics.

E. Mathematics—the Patrimony of the Race. Our discussion has thus led us to ascribe less importance to the role of the individual in the development of mathematics and to give more credit for the creation of this magnificent edifice to the human race as a whole. To be sure, it is always through the gifted individuals that the progress takes place. But no individual is indispensable in this task of furthering mathematical knowledge. The human race produces enough ability of a high degree to make the progress independent of any individual. Albert Einstein said in a press interview: "Individual worship, as I look at it, is always something unjustified. To be sure, nature does distribute her gifts in rich variety among her children. But of those richly gifted ones there are, thank God, many, and I am firmly convinced that most of them lead a quiet unobtrusive existence."

Mathematics is the patrimony of the human race. It is the result of slow and patient labor of countless generations over a period of a great many centuries. Various practical callings have contributed toward this accumulation of mathematical knowledge and have furthered its development in the early and difficult stages. Modern technology provides such stimulation at an ever accelerating pace. The effort which has been expended in erecting the stately and imposing structure which we call mathematics is enormous. But mathematics has repaid the race for this effort. The practical value of mathematics cannot be overemphasized. To those privileged to appreciate the intellectual greatness of mathematics, the contemplation of this grandeur is an endless source of pure joy. The esthetic appeal of mathematics has found its enthusiastic and eloquent exponents. It would be proper to mention here another phase of the merit and value of mathematics to mankind.

The superiority of the human race over all the creatures inhabiting the earth, the reason that mankind is the master of this globe is due primarily to the fact that the experience of each generation does not die with that generation, but is transmitted to the next. This transmission of accumulated experience from generation to generation is the real power of the race, its greatest asset, its most powerful weapon in the conquest of nature, its surest tool in the accumulation of intellectual treasures. Nowhere is this more manifest than in mathematics. The cumulative character of mathematics is really astonishing. There is little in mathematics that ever becomes invalid, and nothing ever gets old. We may have all sorts of non-Euclidean geometries, non-Archimedian geometries, *n*-dimensional geometries, but all this makes the venerable elements of Euclid neither invalid nor obsolete. They remain, graceful and solid, an object of studies as much as ever, all in their own right. This cumulative process, this constant enlargement and perfectability of mathematics is the most precious of its characters, for it has given to mankind the idea of progress, with a clearness and distinctness that nothing else can equal, let alone surpass.

C. J. Keyser in his book *Humanism and Science* [**10**] goes a step farther and points out that the idea of progress suggested by science, and particularly by mathematics, has reflected upon the race itself. It has given mankind the idea that human nature in its turn may be perfected, that with the growth of knowledge and improved living conditions the human race will keep on rising to greater and greater heights on the road toward civilization. Mathematics has given the human race not only the technical tools to bend nature to its uses, not only a great and unequaled storehouse of intellectual beauty and enjoyment, but it also has given mankind a faith in itself and its destinies, hope and courage to carry on this unceasing struggle for a better, more noble, and more beautiful life.

References

1. Law in history, *American Historical Review*, Edward P. Cheyney, vol. 29 (1923–1924), pp. 231–248.

2. *History of Analytic Geometry*, Carl B. Boyer. Scripta Mathematica Studies, New York, 1956,

3. *Outline of the History of Mathematics*, R. C. Archibald. Mathematical Association of America, 1949, p. 99, note 280.

4. Are inventions inevitable?, William F. Ogburn and Dorothy Thomas, *Political Science Quarterly*, vol. 37 (1922), p. 83.

5. *History of Analytic Geometry*, op. cit.

6. The superorganic, A. L. Kroeber, *The American Anthropologist*, vol. 19 (1917), p. 201.

7. The locus of mathematical reality, an anthropological footnote, Leslie A. White, *Philosophy of Science*, vol. 14, no. 4 (October, 1947), p. 298.

8. *Number—The Language of Science*, Tobias Dantzig. First edition, New York, 1930, pp. 195–196.

9. *The Instinct of Workmanship*, Thorstein Veblen New York, 1914.

10. *Humanism and Science*, C. J. Keyser. Columbia University Press, New York, 1931.

Nathan Altshiller was born in Russia in 1881 and received his doctoral degree from the University of Ghent in 1911. He came to the United States, holding positions at Columbia (teaching evening classes), the University of Washington, and the University of Colorado. He joined the University of Oklahoma mathematics department in 1916, where he stayed until his retirement in 1951. On becoming a citizen in 1919 he asked the judge if he could change his name to something more appropriate for a United States citizen, and chose Court. *Mathematics in Fun and in Earnest* was published in 1958. He died in 1968.

He was a geometer, and his *College Geometry* (1925) has been and is widely cited. He also wrote *Modern Pure Solid Geometry* (1935).

4

A Defense of Quadratic Equations

Legislative bodies keep records of their deliberations, for obvious reasons. In the United States, the *Congressional Record* appears every day that the Congress is in session, containing everything that was said in the Senate and the House of Representatives, along with quite a lot that was not said. (Members of Congress can "revise and extend" their remarks, which means that they can insert in the *Record* speeches that they would like to have made, columns from their home-town newspapers, letters from their constituents, or just about anything else.) In the United Kingdom the corresponding journal is *Hansard*.

In the June 26, 2003 issue of *Hansard* appears the record of a speech, actually spoken, by Mr. Tony McWalter, a Labor Member of Parliament for Hemel Hempstead, in defense of quadratic equations. It is a very good defense, which is why it is included here.

Mr. McWalter has a mathematics degree and was for a time the Director of Computing at the University of Hertfordshire, but his specialty is the philosophy of science. Before becoming an MP he was a lecturer in philosophy at the University of Hertfordshire.

To see if the U. S. is the equal of the U. K., I searched the *Congressional Record* for the last ten years, but not one single reference to "quadratic equations" or "quadratic equation" turned up. On the other hand, there were many to be found in *Hansard*. There was one that was not quite to the point:

Sir Nicholas Fairbairn (Perth and Kinross)

Under a Labour Government, a postmistress, a nurse, a postman or a traffic warden who saves a proportion of income for retirement and invests it in the glorious shares of the Scottish power companies would have to pay tax on the income from savings on which he has already paid tax. That is double taxation at its worst. It is a quadratic equation.

It is not a quadratic equation. Sir Nicholas must have had in his mind something about taxation squared, and the association of quadratics with squares caused him to say what he did.

Others were to this sort of point:

Mr. Alex Carlile (Montgomery)

I do not much care—and neither do the people of Wales—which quadratic equation is applied to the funding of local government in Wales. We have heard the Secretary of State's version of the quadratic equation. On the Labour side, we have at least one expert on the quadratic equation from Wrexham. Each could produce their own formula, for what it is worth. What matters much more than the formula is what happens to people.

The Parliamentary Under-Secretary of State for Wales (Mr. Gwilym Jones)

The hon. and learned Member for Montgomery referred to a case concerning Miss J. I am sure that he will understand that I cannot discuss that case now, but if he would care to write to me, I shall certainly pursue it on his behalf. I was drawn by his oratory, which reminded us of his expertise in the courts—after all, he is one of Her Majesty's counsel. How he managed to weave into his argument references to foreplay and quadratic equations was not readily understandable until he told us that an expert in such equations was sitting behind him, the hon. Member for Wrexham (Dr. Marek). I hope that the hon. Gentleman will forgive me, for I had thought that his expertise related to railway timetables.

The point, which is often made in the U. S. as well, is to contrast the arid, theoretical, and irrelevant world of mathematics with the vital and important concerns of people and politicians. It is a natural point for politicians to make, which makes Mr. McWalter's speech even more remarkable.

Quadratic Equations (2003)

26th June 2003

Motion made, and Question proposed, That this House do now adjourn—[Jim Fitzpatrick.]

3:43 pm

Mr. Tony McWalter (Hemel Hempstead): The subject of my debate may ensure that hon. Members will not want to stay for the whole of it, but I dedicate the debate to Sir Nicolas Bevan, who has been the Speaker's Secretary for 10 years. This is the last formal proceeding in the House before he leaves that position. I wanted to pay tribute to him and say that many hon. Members have valued greatly the service that he has given to the House. I know that he will be much missed.

Peter Bottomley (Worthing, West): The hon. Gentleman will have the support of everyone in the House in the tribute that he has paid to the Speaker's Secretary. He might also want to know that the Speaker's Secretary was a classicist and therefore rather better at Latin and Greek than he necessarily was at quadratic equations.

Mr. McWalter: I am grateful to the hon. Gentleman for his intervention and his good wishes to Sir Nicolas. I take the liberty of dedicating a debate to Sir Nicolas because he has shown a strong interest in what one might call the non-conformist debates that have characterized the House from time to time. He has encouraged me to raise with the House the vital philosophical questions that Governments of all persuasions find it too easy to ignore. Despite the mathematical title of the debate, my aim is a philosophical one—it will be an essay in the philosophy of mathematics—and one main objective that I hope to secure is that Sir Nicolas will indeed find it enjoyable.

I put this matter on the agenda today because I have been troubled since the president of a teachers' union suggested a couple of months ago that mathematics might be dropped as a compulsory subject by pupils at the age of 14. Mr. Bladen of the National Association of Schoolmasters and Union of Women Teachers was given a lengthy slot on the "Today" program to present his views. He cited the quadratic equation as an example of the sort of irrelevant topic that pupils study. I had hoped that the Government would make a robust rebuttal, but there was no defense either of mathematics in general or the quadratic equation in particular.

If such assertions are left unrebutted, what was an ignorant suggestion at one time can become received wisdom a very short time later and an article of educational faith a short time after that. I wish to short-circuit that process and provide a rebuttal of that union leader's suggestion. I note that he was a maths teacher, but I do not regard the desire for changes that simply make the teacher's job easier to be in the best union traditions. He was happy to teach maths to those who enjoyed it, but he wanted to stop teaching maths to those who did not. By defending the centrality of the quadratic equation to mathematical education, I hope also to submit some thoughts on what we would be missing if we allowed mathematics to be regarded as a subject of no greater worth than any other subject on the curriculum.

When I proposed this debate, it was arranged that the Minister for School Standards would respond. I submitted a draft to his office, but I note that he has taken flight, so I congratulate my hon. Friend the Minister for Lifelong Learning, Further and Higher Education on having stepped into the breach. I hope that the absence of the Minister for School Standards signifies nothing other than unavailability, as opposed to possible hostility to what I am about to say.

I hope that you will forgive me, madam Deputy Speaker, if I remind the House what an equation is, and then what a quadratic equation is. Of course, we all know that there is a strong appetite for equations in the House—witness the large assembly gathered in the Chamber, as well as the large number of hon. Members who seem to have mastered the mathematical material employing the calculus in the 18 volumes of background support papers for the Chancellor's recent statement on the euro—but it will come as a surprise to hon. Members who are interested in these things to hear that not everyone has an appetite for them. Indeed, it is said that Sir Stephen Hawking was told not to put even a single equation anywhere early in a book of his that sought to popularize science on the grounds that one equation would immediately halve its readership. Apparently, the casual reader flicking through the book in a shop would put it back on the shelf if he or she saw the offending line of print.

What are these equations? From an early stage in primary school, we were given problems such as "If $x + 5 = 7$, what is x?" You will notice, Madam Deputy Speaker, that I am not making the problems too difficult at this stage. Since the time of Descartes, it has been customary to use letters from towards the end of the alphabet for such unknown quantities. Later on, at about the age of 11, we will grapple with or encounter so-called simultaneous equations, where there are two or more unknown quantities and two or more equations.

Even at that stage, many people, whether old or young, feel a sense of bewilderment when such problems are posed, and once the going gets a bit complicated the person who does not want to jump through those hoops is liable to ask, "Why should I bother?" If the education environment is one that says to children, "Study only what interests you", then because the xs and ys look about as boring and detached from reality as anything could be, the pupil is more than pleased when someone in authority says, "If you really don't fancy getting your head around these things, you don't have to." I believe that there is an underlying tension about what we are doing in education, and that the prevailing model is that if someone finds something hard or uninteresting, they are more than welcome to drop it and to move on to something that they find much more tractable and believe, in their minority and youth, to be of much more practical and immediate relevance to their lives.

In that sense, I contend that our educational system has become too focused on working with the current beliefs and enthusiasms of the pupil and insufficiently focused on ignorance. Since education is meant to dispel ignorance—and for all of us appreciating and overcoming our own tendency to ignorance is hard work—an educational model that moves only along the grooves of pupil preference must be deemed too soft. I contend that the soft model should be repudiated and that our model of education should explicitly countenance how important it is for pupils and students to master skills that at first glance seem to them to be strange and uncongenial. Indeed, I might put it more strongly. An idea or a book that seems uncongenial or difficult can, if its subject matter is important, engender those profound changes of attitude that education at its best can precipitate. Education is about climbing mountains, not skipping molehills.

Why should anyone feel passionate about the xs and ys in systems of equations? One answer is this: because if one does not make the effort to see what those xs and ys conceal, one will be cut off from having any real understanding of science. My passion comes from a sense that our society eschews educational difficulty, and hence culturally directs people away from the sciences. What that means—here is the source of my passion—is that in my constituency of Hemel Hempstead, women must wait 18 weeks for a laboratory to process their cervical smear test because many more young people want to work in television than in science, so there are not enough people to work in the laboratory. We have a society that is founded on science, but educationally we provide a university system that offers far more scope for studying the media than for studying physics. I do not wish to engage in the fashionable castigation of media studies or business studies, as many excellent courses go by such names, but where a society provides a very large number of opportunities to study such subjects and a fast-diminishing set of opportunities to study engineering and the mainstream sciences, it makes sense to ask how we have arrived at such a peculiar juncture.

To reflect on that point—and to use an analogy that Sir Nicolas would like—we can observe that a society can regress from being scientifically and technologically cultured to being backward. In the Rome of 800 AD, anyone who wanted to use metal for any purpose would have to find some left by those who lived when the empire was at its zenith. The technical citizens of ancient Rome knew which rocks contained metal ore and developed a furnace technology to liberate the metal from its elemental attendance. Eight hundred years later, such knowledge was entirely lost.

We live in a society that has inherited an extraordinary wealth of knowledge about the world. However, that wealth appears daunting to the pupil or student. To become a scientist appears to require a capacity not only to amass a huge amount of knowledge but to master some ideas, which, at first glance, seem difficult, confusing, remote and mentally too taxing. As David Hume observed, most people have a sufficient disposition toward idleness to want to avoid excessive labor if possible. Consequently, our science-dependent culture is not replenishing the scientific basis that is needed for its continued existence. That neglect has terrible consequences, not only in Hemel Hempstead hospital.

How do quadratic equations relate to all that? First, they are a little more complicated than the linear and simultaneous equations that I mentioned earlier. They have only one unknown expression but they allow it to be raised to a power of 2, for example $x^2 = 4$. Of course, x^2 means that x is multiplied by itself. Another example is $3x^2 + x - 10 = 0$. I suppose that it comes as a shock to find that solving those equations requires some effort.

Even the first one—$x \times x = 4$, what is x?—is not as simple as it appears. There are two solutions: $x = 2$ and $x = -2$. Most people in school learn a general formula to deal with more complicated problems. That is often done without understanding and the world appears to divide into sheep, who do not mind doing that sort of thing and goats, who view it as a pointless game that is less riveting than Snap.

Why should anyone try to understand quadratic equations and the principles that lie behind solving them? They underpin modern science as surely as the smelting methods of the Romans were the key to their building culture. Modern science dawns with the experiments of Galileo. To describe how bodies fall, he knew from Kepler that he had to use the precision of mathematics rather than the imprecise language of Aristotle. The equation that he used for the most fundamental laws of motion was a quadratic equation in time, $s = ut + \frac{1}{2}ft^2$, in which s is distance traveled, u is the initial velocity, f is the accelerating force—usually gravity—and t is time.

To tell students that quadratic equations are beyond them, that they are about nothing and that educated people need have no inkling of what they are is to say that it is all right if they are so ill equipped to understand modern science that they cannot even comprehend its starting point. Those who tell us that we need no familiarity with quadratic equations are telling us to ignore 400 years of intellectual, scientific and technological development. When educators tell us that we should do that, I rejoin that they have a strange view of education, which I should like the Government to repudiate.

If we forget straight lines, the second aspect of the quadratic expression is that it gives us the simplest example of a graph. All admit the utility and importance of that method of presenting information.

If I imagine the simplest quadratic expression, x^2, and I ask of it what values it takes when x assumes different values—when x is 1, x^2 is 1; when x is 2, x^2 is 4, and so on—I get a beautiful elementary curve: a parabola. Galileo used the properties of the parabola to analyze the motion of a falling body. He was able to do so because, long before him, Archimedes had identified some of the properties of the parabola. He knew, for instance, that it was impossible to measure exactly the long side of a unit triangle—a triangle with two sides of length 1, and the longest side, the hypotenuse. It is an extraordinary fact, however, that if such a triangle has a parabola—a curved side—it is possible to measure its area exactly. For example, when the parabola is defined by x^2, when x goes from nought to six, the two straight sides and one curved side will form an area of exactly 72 units.

The mathematical materials of modern science and engineering were laid down by the ancient Greeks, and to tell students that they need not attend to any of these ideas is not merely to deprive them of the ideas that predate Galileo, it is to provide them with an education that neglects entirely the whole post-Hellenic edifice of human scientific culture. The Greeks thought that people were divided into those who could understand at least as far as proposition 47 of book I of Euclid's *Elements*. Anyone who could get beyond that was not an ass. They called that proposition the *pons asinorum*—the asses' bridge. One way of looking at the quadratic equation might be to say that it is the *pons asinorum* of modern science.

I have two further observations on quadratic equations. First, it is powerful educational medicine to come to understand that something that can be expressed very simply can be extraordinarily difficult to solve. Much of modern culture tends the other way. People are presented with enormously difficult problems in politics or economics, for example—I have already mentioned the euro debate—and they assume that such problems have a simple and comprehensible solution. The quadratic equation can teach us to be humble.

Secondly, I have said that to solve such problems one has to make certain moves. In schools, pupils sometimes learn those moves without much understanding of what lies behind them. This is not the place to describe those moves, although I expect that the Minister will be able to remind us of what the generalized solution to a quadratic equation is, because I am sure that his team has equipped him to do so. He probably remembers it anyway from his own schooldays; it is the sort of thing that tends to stick. I am not sure how he is responding to that idea, but I shall persist with the thought.

It is quite extraordinary that one of the outcomes of making efforts to solve these equations is that we seem to have to expand the number system. For example, the solution of the humble-looking equation $2x^2 + 2x + 1 = 0$, a very basic quadratic equation with no hard numbers, seems to require that there be a square root of -1. Since, when we multiply a negative by a negative, we get a positive, it is hard to see how a negative number could have a square root, but the humble quadratic equation suggests that there should be such numbers.

Most people think they know what "number" means; but, in reality, a substantial strand of human intellectual development has involved thinking of how to overcome the limitations of the elementary idea of "number" that we started with, and this rich heritage has been truly a world effort, whether it took place in Iraq, India or China. Knowledge of these things makes people less Anglocentric than they otherwise might be.

Most recently, this bizarre number—the square root of -1—has, since the work of de Broglie in 1923, played a key role in the equations that define quantum theory and which help us to understand our world in its microstructure. I might add to that that the structures that help us understand the other form of equation, simultaneous equations—structures called matrices—are also what are needed in the wave equations of quantum mechanics, since the work of Schrödinger, also in the mid-1920s. If we are to develop nanotechnology, for instance, it will be important that our students are at home with these ideas. Nanotechnology depends upon quantum effects.

I have the honor to serve on the Science and Technology Committee here in the House—you will have probably guessed that by now, Madam Deputy Speaker—and one important aim of that Committee has been to ask the Government to think again about their educational strategy. Some of the concepts and skills that we ask our children and our university students to develop are regarded as "difficult." Mathematics lies in that realm, but so also do other activities such as learning foreign languages, mastering counterpoint and imagining biochemical structures in three dimensions. I submit that such activities, demanding as they do that the students make a real effort to change their perspective, are at the core of education; education as mountain climbing and not molehill jumping.

Students and pupils are told too often that if they find ideas difficult, they can still attain high levels of educational qualification by avoiding such demanding materials. That is actually to do a disservice to those pupils. A key role for education is to help students understand, in all its richness and complexity, the world they have inherited, and perhaps it is also important that they understand the debt they owe to previous generations of many nations and cultures.

A second key role for education in a science-based culture is to equip a significant number of people with the skills to be able to transform that culture for the better. A Government who aspire to have 50 per cent. of school leavers in higher education but who are content for most of those students to have not an inkling of the science and technology that underpins the culture is a Government who are willing to preside over cultural and educational decline, whatever the statistics look like.

Mrs. Eleanor Laing (Epping Forest): Hear, hear.

Mr. McWalter: Oh dear. I would like to have support from elsewhere as well.

Someone who thinks that the quadratic equation is an empty manipulation, devoid of any other significance, is someone who is content with leaving the many in ignorance. I believe also that he or she is also pleading for the lowering of standards. A quadratic equation is not like a bleak room, devoid of furniture, in which one is asked to squat. It is a door to a room full of the unparalleled riches of human intellectual achievement. If you do not go through that door—or if it is said that it is an uninteresting thing to do—much that passes for human wisdom will be forever denied you.

Throughout human history, that door was locked to those of the working classes, to women and to those who come from nations that were enslaved. Now at last we have a society and culture that have made it possible for people on the largest scale to understand at a most fundamental level the culture they have inherited and the debts that they owe to their forebears. Now we have a society in which many citizens can be empowered to understand the natural world.

Siren voices still aver that many cannot cope with quadratic equations and similar structures, and with the worlds that they unlock. It is the Government's job to resist those who would devalue the educational currency in this way. An educational curriculum that is too undemanding cheats those who could have gained understanding, but who are denied that opportunity. Sadly, those who have been so cheated do not even know what it is that they do not know. If real education can be mentally taxing and painful, that is also one of its greatest values. Those who have profited from it are grateful to those who helped them to attain it for the whole of their lives. To deny real education to the many on the dubious ground that they cannot cope with difficulty is to fail to grasp an historic opportunity for human liberation.

Remembering the point about Stephen Hawking, perhaps one day books that feature equations will have their circulation enhanced by that feature. Perhaps we will know that then, we have an educated citizenry.

4:11 pm

The Minister for Lifelong Learning, Further and Higher Education (Alan Johnson): I begin by echoing the comments of my hon. Friend the Member for Hemel Hempstead (Mr. McWalter) on Sir Nicolas Bevan's retirement. I should also like to associate myself with remarks made by Members on both sides of the House throughout the day, and with the sentiments of the well-supported early-day motion. It is not often that a Government representative supports an early-day motion, but I do so on this occasion because it recognizes Sir Nicolas's long and distinguished service. I wish him well in his retirement.

I congratulate my hon. Friend the Member for Hemel Hempstead on securing this debate. He said that he was expecting my hon. Friend the Minister for School Standards to reply, and I should point out that there has been a fierce struggle among the seven Ministers in the Department for Education and Skills in that regard. In the end, my hon. Friend was thought to be far too junior. He has a good career in front of him, but he was part of the 2001 intake and is therefore far too young. So I have the honor to reply to today's debate.

I thank my hon. Friend the Member for Hemel Hempstead for defining what quadratic equations actually are. In fact, my Parliamentary Private Secretary provided me with one, and I shall check it with my hon. Friend afterwards to see whether my Parliamentary Private Secretary will continue to hold his post in future. I hope that I can provide my hon. Friend with a repudiation of the comments, made on the radio a couple of weeks ago, of the trade union leader to whom he referred.

Quadratic equations allow us to analyze the relationships between variable quantities, and they are the tool for understanding variable rates of change. It is in variable rates of change that quadratic equations are seen in economics, science and engineering. Examples of the use of quadratic equations include acceleration, ballistics and financial comparisons. Most drivers would feel capable of working out whether they can overtake the car in front, but do they realize that they are solving a quadratic equation in doing so? I dare say that many do not. In fact, it is claimed that the Babylonians, in 400 BC, were the first to use the notion of quadratic equations in problem solving, although at the time they had no idea what an equation was.

In preparing for this debate, the DFES conducted a straw poll involving a 16-year-old who had just sat maths GCSE, a head of maths and an experienced chemical engineer. The 16-year-old thought that quadratic equations were logical and fairly straightforward because "you substitute stuff into a formula".

He did say, however, his opinion might have been influenced by having a good teacher. The head of maths said that quadratic equations formed an important step in students' ability to solve equations, taking them from simple—one unknown—and simultaneous—two unknowns—and paving the way for more advanced work in mechanics and complex number theory. The engineer said that he did not use quadratic equations now, but had in the past in detailed design applications. Where he works, the chemists use them to explain multiple reactions.

The place of quadratic equations in everyday life is pretty clear, but what are pupils taught at school? The national numeracy strategy has had a significant impact on raising the standards of mathematics in primary schools. Last year's key stage 2 results showed that 73 per cent. of pupils achieved the expected level for their age in mathematics, which is a 14 per cent increase since 1998. We want to build on that impressive record, which is why we have launched "Excellence and Enjoyment—A Strategy for Primary Schools". Our vision for primary education is of excellence and enjoyment at the heart of a broad and rich curriculum.

The key stage 3 national strategy for 11 to 14-year-olds provides a comprehensive professional development program for teachers, new materials and support from expert local consultants. The maths teaching framework emphasizes the development of algebraic reasoning. It encourages pupils to develop an understanding of how algebra is a way of generalizing from arithmetic, and to represent problems and solutions in a variety of forms.

Simple linear equations are taught from key stage 3. In 2002, key stage 3 results stand at their highest ever, with 67 per cent, of pupils achieving level 5 plus in the key stage 3 tests for both maths and science. Quadratic equations, with their more complex parabolic curves, are taught from late key stage 3 or key stage 4. Factorization of algebra is a difficult concept to understand and students need plenty of practice. Teaching students to think logically and to analyze different problems is a very important skill, which is not only transferable to other areas of the curriculum but can be used beyond student life.

[Interruption.] The hon. Member for Epping Forest (Mrs. Laing) from a sedentary position reminds us of that as I speak.

Pupils taking the intermediate tier GCSE mathematics study algebraic manipulation, including the solving of quadratic equations. They are covered in
greater depth by pupils following the higher tier GCSE course and developed
further for those going on to study maths on A and AS-level courses. It is at
that level that students are taught the concepts that they will need, should they
choose to do degrees in maths, sciences, engineering or economics.

There is a shortage of people nationally who can construct these mathematical models and who understand them enough to use them. The fact that an inquiry is taking place into post-14 mathematics is a testament to the importance
of the subject. The aims of the post-14 inquiry, announced in July 2002, are
to "make recommendations on changes to the curriculum, qualifications and
pedagogy for those aged 14 and over in schools, college and higher education institutions to enable those students to acquire mathematical knowledge
and skills necessary to meet the requirements of employers and of further and
higher education".

Professor Adrian Smith, the inquiry chairman, is due to report his findings
this autumn.

The use of information and communications technology in schools and in
teaching and demonstrating mathematics models is helping the understanding
of all learners. With ICT becoming a more integral part of classroom teaching, students can visualize and problem solve in more creative ways. Those,
too, are lifelong skills that can be applied in daily life, not just as a student.
ICT gives students confidence in their abilities and increases their eagerness
to learn. Using graphical calculators to learn about quadratic functions, for
example, helps pupils learn in a more innovative environment.

Several interesting initiatives support teachers in making maths lessons
more challenging and exciting. I should like briefly to describe them. Census
at school can be used in a number of curriculum subjects, particularly maths.
Pupils fill out a questionnaire, see how a census works and are able to compare
their school's results with those of other schools in the UK and elsewhere. That
is particularly helpful for key stage 3 pupils encountering data and learning
how to handle it.

The UK mathematics trust works with secondary school teachers and pupils
to promote mathematics. It encourages all secondary schools to take part in
competitions and events, including the team maths challenge. The trust also
identifies and trains students for the international mathematics olympiad.

The work of the Cambridge university-based millennium maths project allows teachers and students to tap into resources over the internet. Students can
ask university undergraduates for help with mathematical problems. The project also offers tailored and continuous professional development for teachers.

An increasing use of information and communications technology and innovative ways of teaching are both positive steps for the subject, as is the rise in the number of mathematics teachers. The intake to initial teacher training courses in mathematics rose to 1,670 in 2002–03, an increase of 8 per cent., and an overall increase of 28.5 per cent. since 1999–2000. The number of graduates applying to train as teachers of mathematics on postgraduate certificates in education courses in 2003–04 was 35 per cent. higher than for the same period last year. In a recent Ofsted report, it was noted that today's newly qualified teachers are the best trained ever. We need not only to continue to increase the number of teachers, but to support mathematics subject specialism. We want teachers to maintain their enthusiasm for maths and to develop their expertise throughout their careers.

In March, my right hon. friend the Secretary of State for Education and Skills announced that Adrian Smith would advise on options and costs for a national center for excellence in mathematics later this year. The new center should harness all the good work already underway and enable more teachers to tap into the resources and support that they need.

In conclusion, the teaching of quadratic equations, and of the mathematics curriculum overall, is key to a future work force that can develop and use mathematical models in daily life. As research in a book of quotations reveals, Napoleon said: "The advancement and perfection of mathematics are intimately connected with the prosperity of the state."

We recognize the importance of mathematics at all stages of education, and we are committed to ensuring that all young people have the opportunity to acquire the skills that they need—as citizens, and as the mathematicians, scientists and engineers of the future.

Once again, I thank my hon. friend the Member for Hemel Hempstead for securing this debate, and for making such an interesting and entertaining contribution. Question put and agreed to.

Adjourned accordingly at twenty-three minutes past four o'clock.

⁓⟡⟢⁓

Mr. McWalter's dismay at Mrs. Laing's "Hear, hear" is explained by her membership in the Conservative party.

The reply by the Minister for Lifelong Learning, Further and Higher Education was all right, but he fell into a common error when he said "Most drivers would feel capable of working out whether they can overtake the car in front, but do they realize that they are solving a quadratic equation in doing so?" Of course they do not realize that, for the reason that they are not solving one. The alternating current that makes our light bulbs light up flows in sine curves, but we do not need to realize that, or study trigonometry, to be able to flip a light switch on. We do not need to study statics to be able to build a doghouse. Mr. McWalter has more sense than to make such statements.

There is no end to the tug-of-war with teachers on one side, trying to drag students into a world larger than the one they inhabit, and students on the other, digging in their heels and resisting with everlasting cries of "What good is this?" But we must not let the rope go slack, lest we end up in the position of Mr. McWalter's Roman in 800 AD, poking among the ruins of a civilization that he cannot understand.

5

Obtuse Triangles

If you ask someone to draw a triangle at random, what you will almost always get is one that looks very much like an isosceles triangle or a right triangle, like the first two triangles in Figure 0. That is why, when asked to draw a random triangle, I am careful to draw in *AB* and *BC* as in the third triangle in the figure. Even though this means that all my random triangles are obtuse, they are more random than other peoples', as the following paper shows.

The paper is a virtuoso performance by a master, Richard Guy, who is responsible for the Law of Small Numbers among many other things. It is also very funny. Neither of these properties may be apparent to the casual reader. Such can sometimes be the fate of mathematical work, which can contain beauties blushing unseen.

The question the paper considers is, "what proportion of triangles are obtuse?" and gives around a dozen answers. They are thrown off with ease, as a

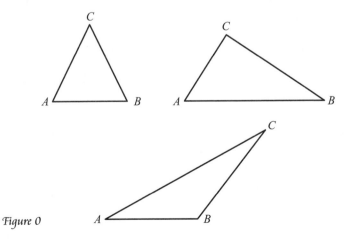

Figure 0

violin virtuoso would handle a passage filled with sixteenth notes, some dou-
ble-stopped. There are *five* proofs that the proportion is 3/4. Ordinary mathematicians find it difficult enough to construct *one* proof of a theorem. Two
might be possible, but five is out of the question. Of course, Gauss proved the
Quadratic Reciprocity Theorem eight different ways, but he was a virtuoso too,
one of the greatest.

It is also funny that there are so many possible answers. The author's way of
presenting them is funny as well. When he writes "The proportion of obtuse-
angled triangles with C in such a region is easily seen to be

$$1 + \frac{e^2}{\sqrt{1-e^2}} - \frac{2}{\pi}\left(\arcsin e + e\sqrt{1-e^2}\right)"$$

he is parodying high-powered mathematical exposition, because it is not easy
to see that, and he piles it on when he tells us that he found it not by using
calculus but by projecting a circle onto an ellipse. Bringing up the GROAT
organization—Greater Rights for Obtuse-Angled Triangles—is funny on two
levels. The first level is the obvious one, and the second is that Professor Guy
knew that he was creating a joke that would, and should, evoke groans, but did
it anyway.

Trying to explain why things are funny is a sure way to suck the humor out
of them, so it is best to stop and let the paper speak for itself. It has values other
than being interesting and entertaining. If it is to be read properly, the reader
has to do some work. When the author says "The fraction of the area giving
an obtuse-angled triangle is $\dfrac{\pi}{2\pi/3+\sqrt{3}}$ or 82.1%." the reader who is reading
properly will stop and go no further until that statement is verified. That cannot
be done by looking at it: machinery must be pulled out and cranks turned to
do things like

$$\int_0^{\pi/3} \left(\tfrac{1}{2}(2\cos\theta)^2 - \tfrac{1}{2}\right)d\theta = \int_0^{\pi/3}\left(2\left(\tfrac{1}{2}+\tfrac{1}{2}\cos 2\theta\right)-\tfrac{1}{2}\right)d\theta$$

$$= \int_0^{\pi/3}\left(\tfrac{1}{2}+\cos 2\theta\right)d\theta = \tfrac{1}{2}\theta + \tfrac{1}{2}\sin 2\theta\Big|_0^{\pi/3}$$

$$= \frac{1}{2}\cdot\frac{\pi}{3}+\frac{1}{2}\cdot\frac{\sqrt{3}}{2}.$$

This too can be thought of as being amusing.

A conclusion, I'm afraid, is that mathematical humor is just too much for
ordinary people, and even for ordinary students of mathematics. That is as may
be, but we should not prevent authors from having fun, especially when they
are virtuosos.

There Are Three Times as Many Obtuse-Angled Triangles as There Are Acute-Angled Ones

Richard K. Guy (1993)

At least! In fact almost all triangles are obtuse-angled, as you will see if you draw one side, *AB* (Figure 1) and then decide where to put the third vertex, *C*. Unless you put it in the shaded region, which constitutes a negligible fraction of the whole plane, you'll get an obtuse triangle. How can we get a more realistic estimate?

If *AB* is to be the shortest side, then *C* must be outside both the circles in Figure 2, and we get a similar result.

If *AB* is to be the middle side, then *C* must be outside one of the circles in Figure 2 and inside the other. The fraction of the area giving an obtuse-angled triangle is

$$\frac{\pi}{2\pi/3 + \sqrt{3}} \quad \text{or} \quad 82.1\%.$$

If *AB* is to be the longest side, then *C* must be inside both circles in Figure 2, and the fraction is

$$\frac{\pi/4}{2\pi/3 - \sqrt{3}/2} \quad \text{or} \quad 63.9\%.$$

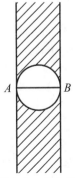

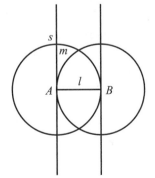

Figure 1. Strip in which *C* forms an acute-angled triangle.

Figure 2. Where *C* makes *AB* shortest (*s*), middle (*m*), or longest (*l*) side.

In practice, we have only a finite region to work in, and we want our triangle to be visible to the naked eye, so a natural region to which to restrict C might be an ellipse with foci A and B (Figure 3).

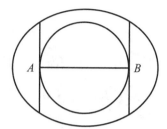

The proportion of obtuse-angled triangles with C in such a region is easily seen to be

$$1+\frac{e^2}{\sqrt{1-e^2}}-\frac{2}{\pi}\left(\arcsin e+e\sqrt{1-e^2}\right)$$

Figure 3. Triangles confined to a golden ellipse.

where e is the eccentricity of the ellipse. As a founding member of SECT[1] let me reassure you that the formula was found by orthogonally projecting a circle and not by using calculus. Of course, if we take the eccentricity to be very small, or very large, then we get an unduly large proportion of obtuse-angled triangles.

But with reasonable numbers, such as $e = 0.618...$, the golden ratio (depicted in Figure 3), or $e = 1/4$, we get the reasonable proportions of 75.24% or 74.96% respectively. However you choose the eccentricity, there'll be more than twice as many obtuse triangles as acute ones.

But most of us work on rectangular sheets of paper. To make the picture "fairly fill the paper" as they used to say in the art exams, draw a 2½-in. line in the middle of an 11-in. × 8½-in. sheet of paper, and the triangle will be obtuse-angled

$$\frac{11\times\left(8\frac{1}{2}-2\frac{1}{2}\right)+\pi\left(\frac{5}{4}\right)^2}{11\times8\frac{1}{2}}=75.83\%$$

of the time. Or, a 6-cm. line in the middle of an A4 sheet (and it's high time we caught up with the rest of the world) yields an obtuse triangle 75.98% of the time.

Another way would be to break a stick, say of unit length, into three pieces, lengths $x + y + z = 1$, and try to make a triangle. Three-quarters of the time it doesn't work, because the pieces must satisfy the triangle inequalities $x + y > z$, etc. This is clear from Figure 4, which is an equilateral triangle of unit height.

A point at distances x, y, z from its sides represents a set of numbers with $x + y + z = 1$. Since x, y, z are positive, we are inside the triangle, and unless we are in the quarter-sized triangle in the middle, one of the triangle inequalities will be violated.

---

[1] The Society for the Elimination of Calculus Teaching, not, as an undergraduate recently suggested, the Society for the Elimination of Calculus Teachers.

But even when we are inside the middle triangle, the angle opposite z will be obtuse if $x^2 + y^2 < z^2 = (1 - x - y)^2$ i.e., if the point is inside the branch of the hyperbola $(1 - x)(1 - y) > 1/2$. The chance that the point is inside one of the three hyperbolas is $9 - 12 \ln 2$ or 68.22%.

"But all this is mere experimental nonsense," I hear you cry, "what about some real mathematics?" Very well, here are no fewer than five (count them) separate genuine proofs.

Proof 1. Every triangle has an **orthocenter**, the common point of the altitudes, that is outside the triangle just if the triangle is obtuse-angled. In any case the vertices and the orthocenter form a set of four *orthocentric points*, each of which is the orthocenter of the other three. Any conic passing through them is a rectangular hyperbola, the locus of whose centers is the nine-point circle..., but I digress! Observe that any three points uniquely determine a fourth and just three of the four triangles thus formed are obtuse-angled.

Proof 2. Every triangle has a **circumcircle**. So choose three points at random on a circle. It's convenient to think of the circumference of the circle as being bent from the stick of unit length that we earlier broke into three pieces. The new interpretation of Figure 4 is that x, y, z are *arcs of the circumcircle* instead of straight lines. A triangle is *always* determined by a point inside the big triangle: The sides are now the associated *chords*. But we know that 3/4 of the time the lengths of the *arcs* don't satisfy the triangle inequality: The three vertices all lie in one half of the circumcircle and the triangle is obtuse-angled.

Proof 3. Every triangle has an **incircle**. So, as in Proof 2, we choose three points randomly on a circle, but this time draw tangents there to form the triangle. But the points must not all lie in the same semicircle, else we would have an **excircle** instead of an incircle. So we are in a similar situation to that of Figure 4: We confine our attention to the middle quarter of Figure 5.

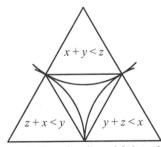

Figure 4. If we break a stick into three parts, x, y, z, then three-quarters of the time we can't make a triangle.

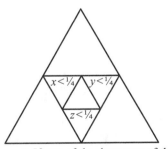

Figure 5. If any of the three arcs of the incircle is less than a quarter, the tangents form an obtuse angle.

Our triangle will be obtuse-angled just if two of the points of tangency lie in the same quadrant of the circle, i.e., just if one of x, y, or $z < 1/4$. These three inequalities are represented by three of the four small triangles in the middle of Figure 5.

Proof 4. Throw down three long straight poles so that they form a triangle. The second pole crosses the first at A as in Figure 6. Of the two supplementary angles at A, call the smaller one θ.

The third pole crosses the first at B (if B falls on the other side of A, rotate the figure through 180 degrees). If the third pole lies in the sectors indicated by the circular arrows, the triangle will be obtuse-angled (at A, B, or C according as the pole lies in the sectors a, b, or c). The fraction of obtuse-angled triangles is

$$\frac{\pi - \theta}{\pi}.$$

We must average this over the interval $0 < \theta < \pi/2$, i.e., $1 - \theta/\pi$ averaged over $0 < \theta/\pi < 1/2$; a linear function that runs from 1 to 1/2; average 3/4.

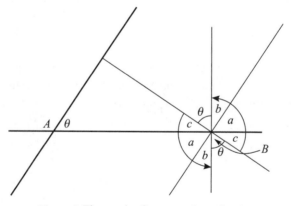

Figure 6. Three poles form a random triangle.

Proof 5. Every triangle has a largest angle. It lies in the range from 60 degrees to 180 degrees. For 3/4 of this range the triangle is obtuse-angled. This proof may be presented in another way. Consider the largest of the three excircles of the triangle. The minor arc of this, between the points of tangency with the arms of the largest angle of the triangle, has length at most 120°. If its length is less than 90°, the largest angle is obtuse, and this occurs 3/4, of the time.

In five well-known geometry books I found figures depicting 146 triangles that purported to be "general" triangles, but only 36 of them were obtuse (often in illustrations of Desargues's theorem, or inversion, or multiplication of complex numbers), almost exactly the reverse proportion of what we now know to be the correct one. You will no doubt immediately join me in supporting the GROAT[2] cause.

Historical Note A good deal of the content of the paper must be scattered about in the literature. Singmaster's *Sources*, under the head "Probability that three lengths form a triangle", traces the "breaking the stick into three pieces" problem at least as far back as Lemoine, and states that there were several later articles based on Lemoine, but neither he nor I have actually seen any of these. He also refers to Fourrey [2], who gives the answer $P = 1/4$ and cites Lemoine [3].

Under the head "Probability that a triangle is acute" he quotes Sylvester and Lewis Carroll. Again he says he hasn't seen these, but gives Miles and Serra [4] as his source for Sylvester. I've located the Sylvester reference [5], a discursive article, with no results. The problem of finding "the chance of three points within a circle or sphere being apices of an acute or obtuse-angled triangle" is attributed to Woolhouse, and his conclusion is that "the form in which ... problems ... originally proposed ... without a specified boundary.... do not admit of a determinate solution." Lewis Carroll [1] states the problem in the form: "Three Points are taken at random on an infinite Plane. Find the chance of their being the vertices of an obtuse-angled Triangle." His solution is essentially that given above under the assumption that AB is the longest side, with the answer

$$\frac{3}{8 - \frac{6\sqrt{3}}{\pi}}.$$

References

1. Lewis Carroll, *Pillow Problems*, 1893. 4th edition (1895) Problem 58, pp. 14, 25, 83–84; reprinted (with *A Tangled Tale*), Dover Publications, New York, 1958.

2. E. Fourrey, *Curiosités Géométriques*, (1st edition Vuibert and Nony, Paris 1907) 4th edition, Vuibert, Paris 1938. Part 3, chap. 1 §5: Application au calcul des probabilités, pp. 360–362. [Break a stick into three pieces.]

3. E. Lemoine, *Bull. Soc. Math. France*, 1872–1873.

4. R. E. Miles and J. Serra, En matière d'introduction, in *Geometrical Probability and Biological Structures*: *Buffon's 200th Anniversary*, Springer Lecture Notes in Biomathematics, 23(1978) 3–28 (esp. p. 18).

---

[2] Greater rights for obtuse-angles triangles.

5. J. J. Sylvester, On a special class of questions in the theory of probabilities, *Birmingham British Association Report* (1865), p. 8. vol. 2 of *The Collected Mathematical Papers of James Joseph Sylvester*, Cambridge, 1908; Chelsea reprint, 1973, 75. pp. 480–481.

The numerical value of the expression found by Lewis Carroll is .639....

Biographical information about Richard Guy can be found at the end of his other item in this collection, number 8, The Law of Small Numbers.

6
A Small Paradox

Why are you so unlucky? Why do you always (it seems) have to wait longer than the average? Is the world malevolently picking on you? Though in our self-pitying moments we may think so, we would be incorrect. The world is not out to get us, it only seems that way because of the inexorable working of mathematics.

When you come to think of it, it is entirely reasonable that we should be in the long line. If there are two lines, one with 99 people and the other with one, the probability that we are in the long one is .99. If the road is clogged with 999 cars at one time and is occupied by only one, zipping along, at another, the chance that we are on the road when it is jammed in .999. It's not that we can't win, it's just that we're not likely to.

This has been noticed more than once and will no doubt be noticed again. Here are two pleasant notes, both explaining the same phenomenon.

Why Your Classes Are Larger than "Average"

David Hemenway (1982)

Most schools advertise their "average class size," yet most students find themselves in larger classes most of the time. Here is a typical example.

In the first quarter of the 1980–81 academic year, 111 courses including tutorials, were given at Harvard School of Public Health. These ranged in size from one student to 229. The **average class size**, from the administration's and professors' perspective, was 14.5. The **expected class size** for a typical student was over 78! This huge discrepancy was due to the existence of a few very large classes. Indeed, only three courses had more than 78 students. One enrolled 105, another 171, and there were 229 in Epidemiology.

Given one class of the size of Epidemiology, an expected class size of approximately 78 for a typical student can be achieved in various ways. Four possible configurations for the rest of the classes are: (I) 450 individual tutorials, (ii) 50 courses of size 10, (iii) 25 courses of size 30, (iv) 25 courses of size 50. The administration's "average class size" for these four cases would be 1.5, 14.3 (close to the advertised figure), 38, and 57 respectively.

The discrepancy between average class size and expected class size for a typical student is explained by a simple computation. Suppose we have a population of M individuals divided into N groups, and we let X_i denote the size of the ith group, $1 \leq I \leq N$. Then the expected number of people in a randomly selected group ("average class size") is given by

$$\bar{X} = \left(\sum X_i\right) / N = M / N,$$

and the expected size of a group containing a randomly selected individual is given by

$$X^* = \sum (X_i / M) X_i = \left(\sum X_i^2\right) / M.$$

Hence

$$X^* - \bar{X} = \left[N \sum_i X_i^2 - \left(\sum X_i\right)^2 \right] / MN$$

$$= \frac{N \sum X_i^2 - \left(\sum X_i^2\right)}{N^2} \cdot \frac{N}{M}$$

$$= \frac{\sigma^2}{\bar{X}},$$

where σ^2 is the variance in group sizes.

The difference between the two means $\bar{X}$ and X^* is directly proportional to the variance in sizes of groups and inversely proportional to average group size. It follows that $X^* \geq \bar{X}$ with equality only when all the groups are the same size.

Here are additional examples from everyday life of the differences between $\bar{X}$ and X^*.

The Nationwide Personal Transportation Survey indicated that average car occupancy ($\overline{X}$) for "home-to-work" trips in metropolitan areas in 1969 was 1.4 people. The table below gives the data.

Number of Occupants	"Home-to-Work" Trips
1	73.5%
2	18.2
3	4.7
4	1.9
5	1.1
6	.5
7	.1

Calculating X^* from these statistics we find that the average number of occupants in the car of a typical commuter was 1.9.

To eliminate most congestion problems in U. S. cities would only require raising the average number of people per car ($\overline{X}$) to 2. This doesn't sound impossible. But suppose this were accomplished by inducing some drivers of single-occupant vehicles to join together in five-person car pools. The percentage of single-occupant cars would need to fall to 58.7%; five-occupant cars would rise to 15.9%. The percentage of people in single-occupant cars would fall below 30%. If X^* is calculated for this situation, one finds that the typical commuter would be in a car carrying more than three people.

I often buy dinner at a fast-food restaurant near my home. Although most customers order "to go," the place is almost always crowded, and I consider it quite a success. One evening about 6:30 I went in and there was no one in line. The manager was serving me, so I asked, "Where is everyone?" "It often gets quiet like this," he said, "even at dinnertime. The customers always seem to come in spurts. Wait fifteen minutes and it will be crowded again." I was surprised that I had never before seen the restaurant so empty. But I probably shouldn't have been. If I am a typical customer, I am much more likely to be there during one of the spurts, so my estimate of the popularity of the restaurant (X^*) is likely to be much greater than its true popularity ($\overline{X}$).

The average number of people at the beach on a typical day will always be less than the average number of people the typical beach-goer finds there. This is because there are lots of people at the beach on a crowded day, but few people are ever there when the beach is practically deserted.

If the waiting time at a health clinic increases with the number of patients, the average waiting time for a typical day will always be less than the average waiting time for a typical patient. This is because there are more patients waiting on those days when the waiting time is especially long.

The expected size of a typical generation will be smaller than the expected number of contemporaries for a randomly chosen individual from one of those generations. Figures for the population density of any region will understate the actual degree of crowding for the average inhabitant.

This Note distinguishes mathematically between two types of means. It does not report any original findings about human behavior. Yet it does indicate something about perceptions —especially my own. I was surprised at the restaurant. I was also surprised when the courses I took in college were larger than advertised. And I was surprised to realize how many commuters had to carpool to reach an average of even two people per car. If you are similarly surprised by any of these observations, your perceptions and perhaps even your behavior may be affected.

Helpful comments were received from Frederick Mosteller and an anonymous referee.

Why is a Restaurant's Business Worse in the Owner's Eyes Than in the Customers'?

Wong Ngai Ying (1987)

Whenever you enter a crowded restaurant, you may think that the owner must be very happy with the business. However, it might not reflect the true picture. The fact is, the time when you enter the restaurant is most probably the time other customers also take their meals. Thus, you may be overlooking the business that occurs during nonbusy hours. Let's look closer at this to see what happens during any period when the restaurant is open for business.

Let the number of customers at time x be $f(x)$ during the business hour $[0, d]$. Then the average number of customers per unit time, as seen by the owner, is

$$\frac{1}{d} \int_0^d f(t)\, dt.$$

On the other hand, if you select a customer at random, the probability that he enters the restaurant in the interval $[x, x + dx]$ is

$$\frac{f(x)}{\int_0^d f(t)\, dt}.$$

Hence, the expected number of customers per unit time as viewed by the customer is

$$\int_0^d \frac{f(x)}{\int_0^d f(t)} f(x)\, dx.$$

The fact that

$$\int_0^d \frac{f^2(x)}{\int_0^d f(t)\, dt}\, dx \ge \frac{1}{d} \int_0^d f(t)\, dt$$

is immediate from the Cauchy inequality, where equality holds if and only if $f(x)$ is a constant function.

Editor's Note. A similar observation was made by Robert Geist in "Perception-Based Configuration Design of Computer Systems," *Information Processing Letters* 18 (1984), where this observation is applied to the design of computing systems which are likely to please users.

The two notes provide yet one more example of the connection between summation and integration. Replace $\sum_{i=1}^n x_i$ with $\int_0^d f(t)\, dt$ and you see that the computations are the same.

They also provide an example of rediscovery. The first note elicited some responses from readers, one pointing out that the result was contained in

Scott Feld and Bernard Grofman, Variation in class size, the class paradox, and some consequences for students, *Research in Higher Education* 6 #3 (1977), pp. 215, 222

and

Scott Feld and Bernard Grofman, Conflict of interest between faculty, students and administrators: consequences of the class size paradox, in *Frontiers of Economics*, Gordon Tullock (ed.), v. 3 (1980).

The inequality dates back to Chebyshev and it, or a generalization thereof, has appeared in at least five places. Details can be found in

Richard A. Brualdi, Comments and complements, *American Mathematical Monthly*, 84 #10 (1977), 804.

But just because something is known, or was known once, does not imply that it is not worth repeating. Every winter there are newspaper headlines saying "Blizzard in North Dakota," but no one writes letters to the editor complaining that the result is known.

7
Applied Mathematics

Curiosity may have killed a cat, and even a human or two, but we are helpless before its power. Human beings want to know what's out there and what's under that rock. We pry, we poke. Our curiosity is next to insatiable. We want to know what is going on, and why.

Mathematicians tend to be curious about numbers. Here is an example of what can happen when a mathematician asks what is behind those numbers on his driver's license.

Assigning Driver's License Numbers

Joseph A. Gallian (1991)

"You know my name look up the number"
— John Lennon and Paul McCartney, *You know my name*,
single, B-side of *Let it be*, March 1970.

Introduction

Among the individual states, a wide variety of methods are used to assign driver's license numbers. The three most common methods, a sequential number, the social security number, and a computer-generated number, are uninteresting mathematically. On the other hand, many states encode data such as month and date of birth, year of birth, and sex in ways that involve elementary mathematics. Seven states go so far as to employ a check digit for possible detection of forgery or errors. Several states assign driver's license numbers by applying complicated hashing functions to the first, middle, and last names and formulas or tables for the month and date of birth. Surprisingly, the assignment of numbers is not always injective. In Michigan, for instance, there are 56 numbers whose inverse image has two or more members. New Jersey incorporates eye color into the number. Some states keep their method confidential. In a few instances, administrators of the license bureaus do not know the method used to assign numbers in their state! In this paper we discuss some of the methods we have uncovered.

Check Digit Schemes in General Use

Schemes for the assignment of identification numbers are extremely varied in methodology and in the information encoded. Most interesting to mathematicians are those that incorporate an extra digit for the detection of errors or fraud. Although the purpose of this paper is to analyze the methods used for driver's license numbers, it is worthwhile to begin with a brief survey of the methods employed to assign check digits to the most ubiquitous numbers in use and to provide a theoretical result that delineates their limitations.

The simplest and least effective method for assigning a check digit is to use the remainder or inverse of the remainder of the identification number modulo some number. For airline tickets, UPS packages, and Federal Express mail the check digit is the identification number modulo 7. An airline ticket number with number 17000459570 is assigned a check digit 3 since $17000459570 \equiv 3 \bmod 7$.

U. S. postal money orders use the remainder modulo 9 while VISA traveler's checks use the inverse of the number modulo 9. Thus, the check digit for the VISA number 1002044679091 is 2 since $1002044679091 \equiv 7 \bmod 9$.

The modulo 7 schemes detect all errors involving a single digit except those where b is substituted for a and $|a - b| = 7$. Likewise, an error of the sort $\ldots a_i \ldots a_j \ldots \rightarrow \ldots a_j \ldots a_i \ldots$ will go undetected if $|a_i - a_j| = 7$ or if 6 divides $j - i$.

The modulo 9 schemes are slightly better at detecting single-digit errors: Only a substitution of a 9 for a 0 or vice versa goes undetected. On the other hand, the only errors of the form $\ldots a_i \ldots a_j \ldots \rightarrow \ldots a_j \ldots a_i \ldots$ that are undetected are those involving the check digit itself. (A quick proof of this is to observe that the residue of a number modulo 9 is the residue of the sum of its digits modulo 9.)

Nearly all methods for assigning a check digit to a string of digits involve a scalar product of two vectors and modular arithmetic. For a string $a_1 a_2 \ldots a_{k-1}$

and a modulus n, many schemes assign a check digit a_k so that

$$(a_1, a_2,\ldots, a_k) \cdot (w_1, w_2,\ldots, w_k) \equiv 0 \bmod n.$$

We call such schemes *linear* and we call the vector $(w_1, w_2,\ldots, w_k)$ the weighting vector. The Universal Product Code (UPC) used on grocery items employs the weighting vector $(3, 1, 3, 1, 3, 1, 3, 1, 3, 1, 3, 1)$ with $n = 10$; the International Standard Book Number (ISBN) utilizes $(10, 9, 8, 7, 6, 5, 4, 3, 2, 1)$ and $n = 11$; banks in the U. S. use $(7, 3, 9, 7, 3, 9, 7, 3, 9)$ with $n = 10$; many Western countries use $(7, 3, 1, 7, 3, 1, \ldots)$ with $n = 10$ to assign check digits to numbers on passports. Notice that the division schemes mentioned at the outset of this section are also linear with weighting vectors of the form $(10^{k-2}, 10^{k-3},\ldots, 10^0, \pm 1)$.

The error-detecting capability of linear schemes is given by the following theorem.

Theorem. *Suppose a number $a_1 a_2 \ldots a_k$ satisfies the condition $(a_1, a_2,\ldots, a_k)$ $\cdot (w_1, w_2,\ldots, w_k) \equiv 0 \bmod n$. Then the single error occasioned by substituting a_i' for a_i is undetectable if and only if $(a_i' - a_i)w_i$ is divisible by n and a sole error of the form $\ldots a_i \ldots a_j \ldots \rightarrow \ldots a_j \ldots a_i \ldots$ is undetectable if and only if $(a_i - a_j)(w_i - w_j)$ is divisible by n.*

Proof. If a_i' is substituted for a_i, then the dot product of the correct number and the incorrect number differ by $(a_i' - a_i)w_i$. Thus, the error is undetected if and only if $(a_i' - a_i)w_i \equiv 0 \bmod n$.

Now consider an error of the form $\ldots a_i \ldots a_j \ldots \rightarrow \ldots a_j \ldots a_i \ldots$. Here the dot product of the correct number and the incorrect number differ by

$$(a_i w_i + a_j w_j) - (a_j w_i + a_i w_j) = (a_i - a_j)(w_i - w_j).$$

The conclusion now follows as before.

Since the most common moduli are 10 and 11, the following corollary is worth mention.

Corollary. *Suppose an identification number $a_1 a_2 \ldots a_k$ satisfies*

$$(a_1, a_2,\ldots, a_k) \cdot (w_1, w_2,\ldots, w_k) \equiv 0 \bmod n$$

where $0 \le a_i < n$ for each i. Then all single-digit errors occurring in the ith position are detectable if and only if w_i is relatively prime to n and all errors of the form $\ldots a_i \ldots a_j \ldots \rightarrow \ldots a_j \ldots a_i \ldots$ are detectable if and only if $w_i - w_j$ is relatively prime to n.

The above theorem verifies our claims about the error-detection capability of the schemes used on money orders and airline tickets. It also explains why the bank and passport schemes will detect some errors of the form $\ldots abc \ldots$

→...*cba*... while the UPC code will detect no such errors. Observe that because 11 is prime the ISBN code detects 100% of all single-digit errors and 100% of all errors involving the interchange of two digits. But there is a price to pay for using the modulus 11: The number a_k needed to satisfy the condition may be 10, which is two digits. In this case, an alphabetic character such as X or A is used or such numbers are not issued. As we will see below there are schemes that use the modulus 11 that do not resort to an alphabetic character, but there is a price to pay for this too: Not all transposition errors are detectable. More information about check digits schemes in use can be found in [1], [2], [3], [4], [6], [7], [8].

Check Digits on Driver's License Numbers

The state of Utah assigns an eight-digit driver's license number in sequential order, say $a_1 a_2 \ldots a_8$, then appends a check digit a_9 using a linear scheme with weighting vector (9, 8, 7, 6, 5, 4, 3, 2, 1) and modulus 10. This method is identical to that used by the American Chemical Society for its chemical registry numbers. Assuming that all errors are equally likely,[1] this method detects 73/81 or 90.1% of all single-digit errors and 100% of all transposition errors (i.e., errors of the form ...*ab*...→...*ba*...).[2]

To verify the single-error detection rate, observe from our theorem that in positions 2, 4, 6, and 8 substitution of b for a will go undetected when $|a - b| = 5$; in position 5, a substitution of b for a will go undetected when a and b have the same parity. Thus in each of positions 2, 4, 6, and 8 there are 10 undetected errors among 90 possible errors while in the fifth position, 40 of the 90 possible errors are undetected. So, in all, 80 of 810 errors are undetected.

Someone working for the Canadian province of Quebec, probably having seen a scheme like the one used by Utah somewhere, came up with the laughable weighting vector (12, 11, 10, 9,..., 2, 1) with modulus 10 to assign a check digit. Of course, any error in the third position is undetectable and weights of 12 and 11 have the same effect as the weights 2 and 1.

Newfoundland uses the weighting vector (1, 2, 3, 4, 5, 6, 7, 8, 1) with modulus 10. This is nearly identical to the Utah scheme except that it will not detect the event that the first and last digit are interchanged.

---

[1] In practice all errors are not equally likely. One study [7, p. 15] revealed that a substitution of a "5" for a "3" was 17 times as likely as a substitution of a "9" for a "1." However, available data are insufficient to assign reliable probabilities to the various error possibilities.

[2] A highly publicized error of this kind recently occurred when Lt. Col. Oliver North gave U.S. Assistant Secretary of State Elliot Abrams an incorrect Swiss bank account number for depositing $10 million for the contras. The correct number begins with "386"; the number North gave to Abrams begins with "368."

Three states use a modified linear scheme with modulus 11. New Mexico and Tennessee append a check digit a_8 to $a_1 a_2 \dots a_7$ as follows: First calculate

$$x = -(a_1, a_2, \dots, a_7) \cdot (2, 7, 6, 5, 4, 3, 2) \bmod 11.$$

If $x = 0$, a_8 is 1; if $x = 10$, $a_8 = 0$; otherwise $a_8 = x$. This method catches 100% of all single-digit errors. Furthermore, the only errors of the form $\dots a_i \dots a_j \dots \rightarrow \dots a_j \dots a_i \dots$ that go undetected are those where $i = 1$ and $j = 7$ (an unlikely error indeed) and some involving the check digits 0 and 1. Assuming that all transposition errors are equally likely,[3] this method detects 98.2% of such errors. The Vermont scheme is the same as the one used by New Mexico except that when $x = 0$, the letter "*A*" is the check. This method, like the ISBN method, yields a 100% detection rate for both single-digit and transposition errors, but utilizes two formats for numbers. Notice that there would be nothing lost if the weighting vector began with 8 instead of 2 and there would be a slight gain since errors of the form $a_1 a_2 \dots a_7 a_8 \rightarrow a_7 a_2 \dots a_1 a_8$ would be detectable.

The state of Washington and the province of Manitoba use a check digit scheme devised by IBM in 1964 to assign a check digit. The license number is a blend of 12 alphabetic and numeric characters. To compute the Washington check digit, alphabetic characters are assigned numeric values as follows: $* \rightarrow 4$, $A \rightarrow 1$, $B \rightarrow 2, \dots, I \rightarrow 9$, $J \rightarrow 1$, $K \rightarrow 2, \dots, R \rightarrow 9$, $S \rightarrow 2$, $T \rightarrow 3, \dots, Z \rightarrow 9$. (Notice the aberration at S.) The 12-character license number, after an alphabetic to numeric conversion, then corresponds to a string of digits $a_1 a_2 \dots a_{12}$ with a_{10} as the check digit calculated as $|a_1 - a_2 + a_3 - a_4 + \dots + a_9 - a_{11} + a12|$ (mod 10). Interestingly, the use of the absolute value actually makes the method nonlinear and reduces the error detection capability of the scheme. It would have been better to use the linear scheme with weighting vector $(1, 9, 1, 9, \dots, 1)$ mod 10.

South Dakota and Saskatchewan employ another nonlinear scheme developed by IBM to assign its check digit. In South Dakota, a six-digit computer-generated string is assigned a check digit as follows. Each of the second, fourth, and sixth digits is multiplied by 2 and the digits of the resulting products are summed (e.g., a 7 yields $1 + 4 = 5$ while a 3 yields 6). This resulting total is then added to the digits in the first, third, and fifth positions. The check digit is the inverse modulo 10 of this tally. (Alternatively, the check digit is

$$\left(10 - \left(\left(\sum_{i \text{ even}} \left(2a_i + \lfloor 2a_i / 10 \rfloor\right) + \sum_{i \text{ odd}} a_i\right)(\bmod 10)\right)\right) \bmod 10.)$$

Thus, the check digit for 263743 is $-(1 + 2 + 1 + 4 + 6 + 2 + 3 + 4) \bmod 10 = 7$.

---

[3] In reality, the likelihood of a transposition error depends on the pair of digits as well as the positions. But as before, reliable data for these occurrences are unavailable.

This method is used by credit card companies, many libraries, and drug stores in the U. S. and by banks in West Germany, although in some instances it is the digits in the odd positions that are multiplied by 2. It detects 100% of all single-digit errors and 97.8% of transposition errors. To see that all single-digit errors are detected, observe that distinct digits contribute distinct values to the sum. To compute the detection rate for errors of the form $...ab... \rightarrow ...ba...$, suppose such an error is undetected. We consider four cases. For simplicity, assume a in the correct number is in position 2, 4 or 6. The alternative case gives the same result.

Case 1. $a, b < 5$
Then $2a + b \equiv 2b + a$ mod 10. Thus $a - b = 0$ and $a = b$.

Case 2. $a < 5, b \geq 5$
Then $2a + b \equiv 2b - 9 + a$ mod 10. It follows that $b - a = 9$ so that $b = 9$ and $a = 0$.

Case 3. $a \geq 5, b < 5$
Then $2a - 9 + b \equiv 2b + a$ mod 10. So, $a - b = 9$ and $a = 9$ and $b = 0$.

Case 4. $a \geq 5, b \geq 5$
Then $2a - 9 + b \equiv 2b - 9 + a$ mod 10. Thus $a - b = 0$ and $a = b$.

So all transposition errors except 09 $\leftrightarrow$ 90 are detected. Since there are 90 possible transposition errors, the error detection rate is 88/90 or 97.8%.

It is worth noting that Gumm [4] has shown that it is not possible to improve upon these rates with any system that uses addition modulo 10 to compute the check digit without utilizing an extra character, as was the case for the New Mexico scheme.

Wisconsin appends a check digit to a 13-digit string. Unfortunately, I have not been able to figure out how this scheme works. I know it isn't linear; for if so, the weighting vector $(w_1, w_2,..., w_{13}, w_{14})$ could be determined by gathering up a large number of valid license numbers to produce a system of linear equations with the w_is as the unknowns. I have done this for modulo 10 and 11 to no avail. To circumvent any peculiarity that might arise involving a check digit of 10 in a modulo 11 scheme (e.g., New Mexico), I avoided numbers with a check digit of 0 or 1.

Encoding Personal Data

Here is the driver's license number of a Wisconsin resident: E 425-7276-9176-07. What information about the holder can you deduce from this number: year of birth, day and month of birth, sex, name? None of these is obvious. Let's go the other direction. I am a resident of Minnesota. I was born on January 5, 1942, and my middle name is Anthony. From this can you deduce my driver's license number?

Eleven states assign their driver's license numbers with hashing functions applied solely to personal data. A good hash function should be fast and minimize collisions (see [5, pp. 506–544] for a detailed discussion of this topic). Of course, there will be occasions when two or more individuals have enough personal data in common that collisions will occur. Most states have a tie-breaking mechanism to handle this situation. Coding license numbers only from personal data enables automobile insurers, government entities, and law enforcement agencies to determine the numbers when necessary.

Washington uses a complicated blend of name, check digit, and codes for the month and date of birth to assign its numbers. This 12-digit identifier consists of the first five letters of the surname; the first and middle initials (* is used when a name has less than five characters, or there is no middle initial); the year of birth subtracted from 100 (we suspect this is done to disguise the year of birth); a check digit; a code for the month of birth; and a code for the day of birth. For instance, Fielding Mellish (no middle name) born on 10/29/42 receives the identifier MELLI F* 587P9. When checked against a file of 1.6 million items, this scheme yielded duplicates at the rate of 0.03% and only one number appeared as many as four times. (Most of the duplications represented twins.) To ensure that the correspondence between individuals and numbers is injective, 17 alternate codes for month and year of birth are available. For example, an S can be used instead of a B for January or a Z instead of a 9 for the year of birth. Interestingly, the check digit is invariant under all alternate coding. The primary code and one alternate for months is given in Table 1 and the code for the days is given in Table 2. Notice the absence of completely predictable patterns.

Illinois, Florida, and Wisconsin encode the surname, first name, middle initial, date of birth, and sex by a quite sophisticated scheme. The first character of the license number is the first character of the name. The next three characters are obtained by applying the "Soundex Coding System" to the surname as follows:

1. Delete all occurrences of h and w.

2. Assign numbers to the remaining letters as follows:

   $b, f, p, v \to 1$ $l \to 4$

   $c, g, j, k, q, s, x, z \to 2$ $m, n \to 5$

   $d, t \to 3$ $r \to 6$

   (No values are assigned to $a, e, i, o, u,$ and y.)

Table 1. Washington code for months.

Months	Codes	Alternate Codes
January	B	S
February	C	T
March	D	U
April	J	1
May	K	2
June	L	3
July	M	4
August	N	5
September	O	6
October	P	7
November	Q	8
December	R	9

Table 2. Washington code for days.

1 - A	7 - G	13 - L	19 - R	25 - 5	31 - U
2 - B	8 - H	14 - M	20 - 0	26 - 6	
3 - C	9 - Z	15 - N	21 - 1	27 - 7	
4 - D	10 - S	16 - W	22 - 2	28 - 8	
5 - E	11 - J	17 - P	23 - 3	29 - 9	
6 - F	12 - K	18 - Q	24 - 4	30 - T	

3. If two or more letters with the same numeric value are adjacent, omit all but the first. (Here *a, e, i, o, u,* and *y* act as separators.) For example, Schworer becomes Sorerr and Hughgill becomes Ugil.

4. Delete the first character of the original name if still present.

5. Delete all occurrences of *a, e, i, o, u,* and *y.*

6. Use the first three digits corresponding to the remaining letters; append trailing zeros if less than three letters remain.

Here are some examples: Schworer → S-660; Hughgill → H-240; Skow → S-000; Sachs → S-200; Lennon → L-550; McCartney → M-263.

We parenthetically remark that the Soundex System was designed so that likely misspellings of a name would nevertheless result in the correct coding of the name. For example, frequent misspellings of my name are: Gallion, Gillian, Galian, Galion, Gilliam, Gallahan, and Galliam. Observe that all of these yield the same coding as Gallian. We also mention that the above method differs somewhat from the system called Soundex by Knuth in [5, p. 392].

The next three digits are determined by summing numbers that correspond to the first name and middle initial. The scheme for doing this begins with the block 000 for the letter *A* and makes jumps of 20 for especially common names and each subsequent letter of the alphabet. A small portion of this scheme is given in Table 3. The values assigned to the middle initial are given in Table 4.

Table 3. Illinois, Florida, Wisconsin given name or first initial code.

000 - A
020 - Albert, Alice
040 - Ann, Anna, Anne, Annie, Arthur
060 - B
080 - Bernard, Bette, Bettie, Betty
100 - C

Table 4. Illinois, Florida, Wisconsin middle initial code.

0 - none	10 - J
1 - A	11 - K
2 - B	12 - L
3 - C	13 - M
4 - D	14 - N ,O
5 - E	15 - P, Q
6 - F	16 - R
7 - G	17 - S
8 - H	18 - T, U, V
9 - I	19 - W, X, Y, Z

So Aaron G. Schlecker would be coded as S426-007 (S426 from Schlecker; 000 for Aaron + 7 for "G"), while Anne P. Schlecker would be coded as S426-055.

The last five digits of Illinois and Florida numbers capture the year and date of birth as well as the sex. In Illinois, each day of the year is assigned a three-digit number in sequence beginning with 001 for January 1. However, each month is assumed to have 31 days. Thus, March 1 is given 063. These numbers are then used to identify the month and day of birth of male drivers. For females, the scheme is identical except January 1 begins with 601. The last two digits of the year of birth, separated by a dash (probably for camouflage), are listed in the 5th and 4th positions from the end of the driver's license number. Thus, a male born on July 18, 1942, would have the last five digits 4-2204 while a female born on the same day would have 4-2804. When necessary, Illinois adds an extra character to avoid duplications. Among the 9,397,518 licenses on file on January 1, 1987, this occurred in 14,856 instances. Of these, 55 numbers corresponded to three individuals (excluding the extra digit). No number corresponded to more than three people.

The scheme to identify birthdate and sex in Florida is the same as Illinois except each month is assumed to have 40 days and 500 is added for women. For example, the five digits 49583 belong to a woman born on March 3, 1949.

Wisconsin employs the same scheme as Florida to generate the first 12 of their 14 characters. The thirteenth character is an integer issued sequentially beginning with 0 to people who share the same first 12 characters. The fourteenth character is a check digit.

A Missouri driver's license number has 16 characters. The first 13 characters are obtained by applying a hashing function to the first five letters of the surname, the first three letters of the first name and the middle initial. The final three characters are a function of the month and day of birth and sex. For a male born in month m and day d the three digits are $63m + 2d$. For a female, the corresponding formula is $63m + 2d + 1$. Thus the number of a woman born on March 4 has the final three characters 198. To avoid duplications, Missouri assigns a 17th character. Among the first 3,921,922 numbers issued, 31,719 have a 17th character.

Last, we discuss the scheme employed by Minnesota, Michigan, and Maryland. The number is a function of last name, first name, middle name, month and date of birth. The first four characters are determined by the Soundex System, as was the case for Illinois, Florida, and Wisconsin. The first and middle names account for the next six characters and the same algorithm is applied to both names. In the majority of cases the first two characters of the name determine the desired three digits for each name (see Table 5 for a sample); for common pairs of leading letters such as *Al* or *Ja*, the third letter is invoked (see Table 6); 11 three-digit numbers are uniquely assigned to the 11 most popular names (e.g., 189 Edward; 210 Elizabeth).

The final three digits are based on month and date of birth (but not year). Each day of the year is assigned a three-digit number in a monotonically increasing fash-

Table 5. Minnesota, Michigan, Maryland code for first and middle names beginning with *A* except *Al*.

		A	027		
Aa	028	Aj	037	As	072
Ab	029	Ak	038	At	073
Ac	030	Al	—	Au	074
Ad	031	Am	066	Av	075
Ae	032	An	067	Aw	076
Af	033	Ao	068	Ax	077
Ag	034	Ap	069	Ay	078
Ah	035	Aq	070	Az	079
Ai	036	Ar	071		

ion. Although the usual pattern is to alternate increments of 3 and 2, there are numerous seemingly random increments at unpredictable dates. The month of March illustrates this behavior well. Notice from Table 7 that March 1 is assigned 159. Subsequent days are assigned values by increments of 3 and 2 in alternating fashion until March 8. Then there is an increment of 5. Notice the jump of 20 between March 19 and March 20.

These gaps serve a practical purpose. In the event that there are two or more individuals born on the same month and date and with names so similar that the hashing function does not distinguish between them (e.g., Jill Paula Smith and Jimmy Paul Smythe), the first person who applies for a license is assigned the number given by the algorithm while the second person is assigned the next higher number thereby using one of the num-

Table 6. Minnesota, Michigan, Maryland code for first and middle names beginning with *Al*.

		Al	039		
Ala	040	Alj	049	Als	058
Alb	041	Alk	050	Alt	059
Alc	042	All	051	Alu	060
Ald	043	Alm	052	Alv	061
Ale	044	Aln	053	Alw	062
Alf	045	Alo	054	Alx	063
Alg	046	Alp	055	Aly	064
Alh	047	Alq	056	Alz	065
Ali	048	Alr	057		

Table 7. Minnesota, Michigan, Maryland code for dates in March.

		March - 158			
1	- 159	11	- 187	21	- 229
2	- 162	12	- 189	22	- 232
3	- 164	13	- 192	23	- 234
4	- 167	14	- 194	24	- 237
5	- 169	15	- 197	25	- 239
6	- 172	16	- 199	26	- 242
7	- 174	17	- 202	27	- 244
8	- 177	18	- 204	28	- 247
9	- 182	19	- 207	29	- 249
10	- 184	20	- 227	30	- 252
				31	- 254

bers in the gap for birthdays. For example, if Jill Paula Smith is born on March 2 and is the first to receive the combination S530-441-675-162 as determined by the algorithm, then the next person who yields the same number is assigned S530-441-675-163 instead. Once all of the higher numbers in a gap have been assigned, lower numbers are used. Thus the third applicant with a name yielding the combination S530-441-675 born on March 2 would be assigned the last three digits 161. As of 1984, this scheme had not yielded any duplications among 4,468,080 people in Maryland while of Michigan's 6,332,878 drivers by 1987 there are 56 that have a number not uniquely their own. In fact, Michigan has two numbers that are each shared by four individuals and three that are each shared by three individuals. A common cause of duplication is the custom of naming a son after the father. When both share the same birthday a duplication occurs.

Summary

Table 8 summarizes the information the author has discovered about the coding of driver's license numbers. Unfortunately our knowledge is incomplete. Several states (e.g., Florida, New York, Minnesota, Missouri, Wisconsin) keep their methods confidential. In some of these cases we were able to determine the coding scheme by examining data. A question mark after the letter X indicates the corresponding item is used in the coding, but we do not know the method involved. The expression (A) after an X indicates that the corresponding item is part of a scheme that is an alternative to the social security number.

Table 8. Summary of Schemes for Assigning Driver's License Numbers

State	Social Security Number	Comp. or Sequential Number	Check Digit	Last Name Coded	First Name Coded	Middle Name Coded	Year of Birth Coded	Month of Birth Coded	Day of Birth Coded	Sex Coded
Alabama		X								
Alaska		X								
Arizona	X									
Arkansas	X	X(A)								
California		X								
Colorado		X								
Connecticut		X						X		
Delaware		X					X			
Florida				X	X	X	X	X	X	X
Georgia	X	X(A)								
Hawaii	X									
Idaho	X	X(A)								
Illinois				X	X	X	X	X	X	X
Iowa	X									
Kansas		X								
Kentucky	X									
Louisiana		X								
Maine		X		X	X		X	X	X	
Maryland				X	X	X		X	X	
Massachusetts	X									
Michigan				X	X	X		X	X	
Minnesota				X	X	X		X	X	
Mississippi	X									
Missouri				X(?)	X(?)	X(?)		X	X	X

State	Social Security Number	Comp. or Sequential Number	Check Digit	Last Name Coded	First Name Coded	Middle Name Coded	Year of Birth Coded	Month of Birth Coded	Day of Birth Coded	Sex Coded
Montana	X				X (A)		X(A)	X (A)	X (A)	X (A)
Nebraska		X								
Nevada	X	X(A)								
New Hampshire				X	X		X	X	X	
New Jersey				X	X	X	X	X		X
New Mexico		X	mod 11							
New York				X(?)	X(?)	X(?)	X	X(?)	X(?)	?
North Carolina		X								
North Dakota	X	X (A)								
Ohio		X								
Oklahoma	X	X(A)								
Oregon		X								
Pennsylvania		X								
Rhode Island		X								
South Carolina		X								
South Dakota		X	mod 10				X	X		
Tennessee		X	mod 11							
Texas		X								
Utah		X	mod 10							
Vermont		X	mod 11							
Virginia	X	X(A)								
Washington			mod 10	X	X	X	X	X	X	
West Virginia		X								
Wisconsin			X(?)	X	X	X		X	X	X
Wyoming		X	X(?)							

References

1. Steve Connor, The invisible border guard, *New Scientist* (Jan. 5 1984), 9–14.

2. Joseph A. Gallian, The zip code bar code, *The UMAP Journal* 7 (1986), 191–194.

3. Joseph A. Gallian and Steven Winters, Modular arithmetic in the marketplace, *American Mathematical Monthly* 95 (1988), 548–551.

4. H. Peter Gumm, A new class of check digit methods for arbitrary number systems, *IEEE Transactions on Information Theory* 31 (1985), 102–105.

5. Donald E. Knuth, *The Art of Computer Programming*, vol. 3, Addison-Wesley, Reading, MA, 1973.

6. Philip M. Tuchinsky, International standard book numbers, *The UMAP Journal* 5 (1985), 41–54.

7. J. Verhoeff, *Error Detecting Decimal Codes*, Mathematical Centrum, Amsterdam, 1969.

8. E. F. Wood, Self-checking codes—an application of modular arithmetic, *Mathematics Teacher* 80 (1987), 312–316.

Addendum

Subsequent to publication of this article the author was able to use samples to determine the method used by Missouri and New York (see *UMAP Journal*, 13 (1992) 37–42) as well as the check digit scheme used by Wisconsin. Wisconsin assigns a numerical value for the initial letter the same way Washington does and uses the South Dakota method for calculating the check digit.

This selection was titled "Applied Mathematics" because the work in it had an application, a useful one outside of mathematics. A university was conducting a medical study extending over several years and needed to get in touch with those participating in it. Most of the people could be reached, but as always there were those who had moved, leaving no forwarding address. Even with a forwarding address, the Post Office stops sending mail on after a year. So, how to locate the missing persons?

The researchers had the idea of asking the state automobile license bureau for their addresses. The bureau would not give an address for a name; the rules, or the law, would not permit it. But, the researchers were told, if they supplied the person's driver's license number, the license bureau would tell them the address of the person who held it—that would not violate the rules. (There are no doubt good reasons for both of those rules.) The researchers went to Professor Gallian, who turned the names into license numbers. They sent them in, and got the addresses that they needed. Let no one say that mathematical curiosity that seems idle does not have use.

Joseph Gallian is a professor of mathematics at the University of Minnesota Duluth. He is the Co-director of Project NExT. He is also the author of a popular abstract algebra text and of many papers, several of which have won awards. He has conducted a very successful Research Experiences for Undergraduates program for more than twenty years and is, as the quote at the start of his paper indicates, a Beatles scholar.

8

The Law of Small Numbers

Christopher Columbus sailed with three ships, there are three Rs, three blind mice, three Fates (Clotho, Lachesis, and Atropos), three Musketeers, three kingdoms (animal, vegetable, and mineral), three Furies (Alecto, Megaera, and Tisiphone), three bags of wool (one for my master, one for my dame, and one for the little boy who lives down the lane), three dimensions, and Old King Cole called for his fiddlers three. Well! There surely must be something special about three.

There is. It is a small integer, and it should be no surprise that many sets of things have cardinality 3. Quite a few have cardinality 7 and, come to think of it, there are many 4s too. Though some people see mystical significance in all those 3s (or 7s, or 4s), all that is operating is Richard Guy's Law of Small Numbers, a universal Law that deserves wide recognition. Here is its original statement.

The Strong Law of Small Numbers

Richard K. Guy (1988)

This article is in two parts, the first of which is a do-it-yourself operation, in which I'll show you 35 examples of patterns that seem to appear when we look at several small values of *n*, in various problems whose answers depend on *n*. The question will be, in each case: do you think that the pattern persists for all

n, or do you believe that it is a figment of the smallness of the values of n that are worked out in the examples?

Caution: examples of both kinds appear; they are not all figments!

In the second part I'll give you the answers, insofar as I know them, together with references.

Try keeping a scorecard: for each example, enter your opinion as to whether the observed pattern is known to continue, known not to continue, or not known at all.

This first part contains no information; rather it contains a good deal of disinformation. The first part contains one theorem:

> You can't tell by looking.

It has wide application, outside mathematics as well as within. It will be proved by intimidation.

Here are some well-known examples to get you started.

Example 1. The numbers $2^{2^0}+1=3$, $2^{2^1}+1=5$, $2^{2^2}+1=17$, $2^{2^3}+1=257$, $2^{2^4}+1=65537$ are primes.

Example 2. The number 2^n-1 can't be prime unless n is prime, but $2^2-1=3$, $2^3-1=7$, $2^5-1=31$, $2^7-1=127$, are primes.

Example 3. Apart from 2, the oddest prime, all primes are either of shape $4k-1$, or of shape $4k+1$. In any interval $[1, n]$, the former are at least as numerous as the latter ($4k-1$ wins the "prime number race"):

$4k-1$ 3   7   11   19   23   31   43   47   59   67   71   79   83   103   107   127   131   139

$4k+1$ 5   13   17   29   37   41   53   61   73   89   97   101   109   113   137   149

Example 4. Pick several numbers at random (it suffices just to look at odd ones). Estimate the probability that a number has more divisors of shape $4k-1$, than it does of shape $4k+1$. For example, 21 has two of the first kind (3 and 7) and two of the second (1 and 21), while 25 has all three (1, 5, 25) of the second kind.

Example 5. The five circles of Fig. 1 have $n=1, 2, 3, 4, 5$ points on them. These points are in general position, in the sense that no three of the $\binom{n}{2}$ chords joining them are concurrent. Count the numbers of regions into which the chords partition each circle.

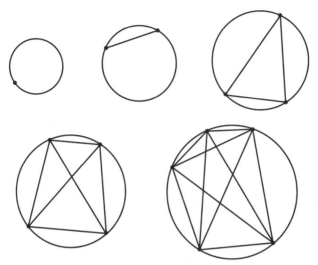

Figure 1. How many regions in each of these circles?

I've been trying to formulate the Strong Law of Small Numbers for many years [9]. The best I can do so far is

> There aren't enough
> small numbers to meet the
> many demands made of them.

It is the enemy of mathematical discovery. When you notice a mathematical pattern, how do you know it's for real?

> Superficial similarities
> spawn spurious statements.

> Capricious coincidences
> cause careless conjectures.

On the other hand, the Strong Law often works the other way:

> Early exceptions
> eclipse eventual essentials.

> Initial irregularities
> inhibit incisive intuition.

Here are some misleading facts about small numbers:

Ten per cent of the first hundred numbers are perfect squares.

A quarter of the numbers less than 100 are primes.

Except for 6, all numbers less than 10 are prime powers.

Half the numbers less than 10 are Fibonacci numbers

$$0, 1, 1, 2, 3, 5, 8,\ldots$$

and alternate Fibonacci numbers, 1, 2, 5,... are both Bell numbers and Catalan numbers.

Example 6. The numbers 31, 331, 3331, 33331, 333331, 3333331, are each prime.

Example 7. The alternating sums of factorials,

$$3! - 2! + 1! = 5$$
$$4! - 3! + 2! - 1! = 19$$
$$5! - 4! + 3! - 2! + 1! = 101$$
$$6! - 5! + 4! - 3! + 2! - 1! = 619$$
$$7! - 6! + 5! - 4! + 3! - 2! + 1! = 4421$$
$$8! - 7! + 6! - 5! + 4! - 3! + 2! - 1! = 35899$$

are each prime.

Example 8. In the table

Row																													Count
Row 1												1	1																2
Row 2											1		2		1														3
Row 3										1		3	2	3			1												5
Row 4								1		4	3		2		3	4		1											7
Row 5						1		5	4		3	5	2	5	3		4	5	1										11
Row 6					1		6	5	4		3	5	2	5	3		4	5	6	1									13
Row 7			1	7	6	5		4	7	3		5	7	2	7	5	3	7	4		5	6	7	1					19
Row 8	1		8	7	6	5	4	7	3	8	5	7		2		7	5	8	3	7	4	5	6	7	8		1		23
Row 9	1	9	8	7	6	5	9	4	7	3	8	5	7	9	2	9	7	5	8	3	7	4	9	5	6	7	8	9 1	29

row n is obtained from row $n - 1$ by inserting n between each pair of consecutive numbers which add to n. The number of numbers in each row is shown on the right. Each is prime.

Example 9. Is there a prime of shape $7013 \times 2^n + 1$?

Example 10. Are all the numbers $78557 \times 2^n + 1$ composite?

Example 11. When you use Euclid's method to show that there are unboundedly many primes:

$$2 + 1 = 3$$
$$(2 \times 3) + 1 = 7$$

$$(2\times3\times5)+1=31$$
$$(2\times3\times5\times7)+1=211$$
$$(2\times3\times5\times7\times11)+1=2311$$

you don't always get primes:

$$(2\times3\times5\times7\times11\times13)+1=30031=59\cdot509$$
$$(2\times3\times5\times7\times11\times13\times17)+1=510511=19\cdot97\cdot277$$
$$(2\times3\times5\times7\times11\times13\times17\times19)+1=9699691=347\cdot27953$$

but if you go to the *next* prime, its difference from the product is always a prime:

$$5-2=3$$
$$11-(2\times3)=5$$
$$37-(2\times3\times5)=7$$
$$223-(2\times3\times5\times7)=13$$
$$2333-(2\times3\times5\times7\times11)=23$$
$$30047-(2\times3\times5\times7\times11\times13)=17$$
$$510529-(2\times3\times5\times7\times11\times13\times17)=19$$
$$9699713-(2\times3\times5\times7\times11\times13\times17\times19)=23$$

Example 12. From the sequence of primes, form the first differences, then the absolute values of the second, third, fourth ,... differences:

2	3	5	7	11	13	17	19	23	29	31	37	41	43	47	53	59	61	67
	1	2	2	4	2	4	2	4	6	2	6	4	2	4	6	6	2	6
		1	0	2	2	2	2	2	2	4	4	2	2	2	2	0	4	4
			1	2	0	0	0	0	0	2	0	2	0	0	0	2	4	0
				1	2	0	0	0	0	2	2	2	2	0	0	2	2	4
					1	2	0	0	0	2	0	0	0	2	0	2	0	2
						1	2	0	0	2	2	0	0	2	2	2	2	2
							1	2	0	2	0	2	0	2	0	0	0	0
								1	2	2	2	2	2	2	2	0	0	0
									1	0	0	0	0	0	0	2	0	0
										1	0	0	0	0	0	2	2	0
											1	0	0	0	0	2	0	2

Is the first term in each sequence of differences always 1?

Example 13. 2^n is never congruent to 1 (mod n) for $n>1$. 2^n is congruent to 2 (mod n) whenever n is prime, and occasionally when it isn't ($n=341,561,...$). Is 2^n ever congruent to 3 (mod n) for $n>1$?

Example 14. The good approximations to $5^{1/5}$, namely, the convergents to

$$1 + \cfrac{1}{2+} \cfrac{1}{1+} \cfrac{1}{1+} \cfrac{1}{1+} \cfrac{1}{2+} \cdots \quad \text{are} \quad \frac{1}{1}, \frac{3}{2}, \frac{4}{3}, \frac{7}{5}, \frac{11}{8}, \frac{29}{21}, \dots$$

which have Fibonacci numbers for denominators and Lucas numbers for numerators.

Example 15.

$$(x+y)^3 = x^3 + y^3 + 3xy(x^2 + xy + y^2)^0$$
$$(x+y)^5 = x^5 + y^5 + 5xy(x^2 + xy + y^2)^1$$
$$(x+y)^7 = x^7 + y^7 + 7xy(x^2 + xy + y^2)^2.$$

Example 16. The sequence of *hex numbers* (so named to distinguish them from the hexagonal numbers, $n(2n-1)$) are depicted in Fig. 2. The partial sums of this sequence, 1, 8, 27, 64, 125, appear to be perfect cubes.

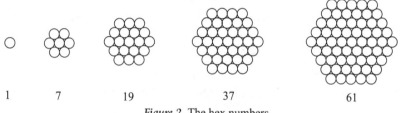

1   7   19   37   61

Figure 2. The hex numbers.

Example 17. Write down the positive integers, delete every second, and form the partial sums of those remaining:

1	2̸	3	4̸	5	6̸	7	8̸	9	1̸0̸	11
1		4		9		16		25		36

Example 18. As before, but delete every third, then delete every second partial sum:

1	2	3̸	4	5	6̸	7	8	9̸	10	11	1̸2̸	13	14	1̸5̸	16
1	3̸		7	1̸2̸		19	2̸7̸		37	4̸8̸		61	7̸5̸		91
1			8			27			64			125			216

Example 19. Again, but delete every fourth, then every third partial sum, then every second of their partial sums:

1	2	3	~~4~~	5	6	7	~~8~~	9	10	11	~~12~~	13	14	15	~~16~~	17
1	3	~~6~~		11	17	~~24~~		33	43	~~54~~		67	81	~~96~~		113
1	~~4~~			15	~~32~~			65	~~108~~			175	~~256~~			369
1				16				81				256				625

Example 20. Again, but circle the first number of the sequence, delete the second after that, the third after that, and so on. Form the partial sums and repeat:

(1)	2	~~3~~	4	5	~~6~~	7	8	9	~~10~~	11	12	13	14	~~15~~	16	17	18	19	20	~~21~~
	(2)		6	~~11~~		18	26	~~35~~		46	58	71	~~85~~		101	118	136	155	~~175~~	
			(6)			24	~~50~~			96	154	~~225~~			326	444	580	~~735~~		
						(24)				120	~~274~~				600	1044	~~1624~~			
										(120)					720	~~1764~~				
															(720)					

Example 21. Write down the odd numbers starting with 43. Circle 43, delete one number, circle 47, delete two numbers, circle 53, delete three numbers, circle 61, and so on. The circled numbers are prime (Fig. 3).

(43)	~~45~~	(47)	~~49~~	~~51~~	(53)	~~55~~	~~57~~	~~59~~	(61)	~~63~~	~~65~~	~~67~~	~~69~~	(71)	~~73~~	~~75~~	~~77~~
~~79~~	~~81~~	(83)	~~85~~	~~87~~	~~89~~	~~91~~	~~93~~	~~95~~	(97)	~~99~~	~~101~~	~~103~~	~~105~~	~~107~~	~~109~~	~~111~~	(113)
~~115~~	~~117~~	~~119~~	~~121~~	~~123~~	~~125~~	~~127~~	~~129~~	(131)	~~133~~	~~135~~	~~137~~	~~139~~	~~141~~	~~143~~	~~145~~	~~147~~	~~149~~
(151)	/	/	/	/	/	/	/	/	/	/	(173)	/	/	/	/	/	/
/	/	/	/	/	(197)	/	/	/	/	/	/	/	/	/	/	/	/
(223)	/	/	/	/	/	/	/	/	/	/	/	/	/	(251)	/	/	/
/	/	/	/	/	/	/	/	/	/	/	(281)	/	/	/	/	/	/
/	/	/	/	/	/	/	/	/	(313)	/	/	/	/	/	/	/	/
/	/	/	/	/	/	/	/	(347)	/	/	/	/	/	/	/	/	/
/	/	/	/	/	/	/	/	(383)	/	/	/	/	/	/	/	/	/
/	/	/	/	/	/	/	/	/	(421)	/	/	/	/	/	/	/	/
/	/	/	/	/	/	/	/	/	/	/	(461)	/	/	/	/	/	/
/	/	/	/	/	/	/	/	/	/	/	/	/	/	(503)	/	/	/
/	/	/	/	/	/	/	/	/	/	/	/	/	/	/	/	/	/
(547)	/	/	/	/	/	/	/	/	/	/	/	/	/	/	/	/	/
/	/	/	/	/	(593)	/	/	/	/	/	/	/	/	/	/	/	/
/	/	/	/	/	/	/	/	/	/	/	(641)	/	/	/	/	/	/

Figure 3. Parabolas of primes remain.

Example 22. In Table 1 the odd prime values of $n^4 + 1$ and of $17 \times 2^n - 1$ are printed in bold. They occur simultaneously for $n = 2, 4, 6, 16, 20$.

Table 1

n	$n^4 + 1$	$17 \times 2^n - 1$	n	$n^4 + 1$	$17 \times 2^n - 1$
0	1	$16 = 2^4$	12	$20737 = 89 \times$	$69631 = 179 \times$
1	2	$33 = 3 \times 11$	13	$28562 = 2 \times$	$139263 = 3 \times$
2	**17**	**67**	14	$38417 = 41 \times$	$278527 = 233 \times$
3	$82 = 2 \times 41$	$135 = 3^3 \times$	15	$50626 = 2 \times$	$557055 = 3^2 \times$
4	**257**	**271**	16	**65537**	**1114111**
5	$626 = 2 \times$	$543 = 3 \times$	17	$83522 \times$	$2228223 = 3 \times$
6	**1297**	**1087**	18	$104977 = 133 \times$	$4456447 = 59 \times$
7	$2402 = 2 \times$	$2175 = 3\_$	19	$130322 = 2 \times$	$8912895 = 3 \times$
8	$4097 = 17 \times$	$4351 = 19 \times$	20	**160001**	**17825791**
9	$6562 = 2 \times$	$8703 = 3^2\_$	21	$194482 = 2 \times$	$35651583 = 3^4 \times$
10	$10001 = 73 \times$	$17407 = 13^2 \times$	22	$234257 = 73 \times$	$71303167 = 13 \times$
11	$14642 = 2 \times$	$34815 = 3 \times$	23	$279842 = 2 \times$	$142606335 = 3 \times$

Example 23. In Table 2 the prime values of $21 \times 2^n - 1$ and of $7 \times 4^n + 1$ are printed in bold. They occur simultaneously for $n = 1, 2, 3, 7, 10, 13$.

Table 2

n	$21 \times 2^n - 1$	$7 \times 4^n + 1$	n	$21 \times 2^n - 1$	$7 \times 4^n + 1$
0	$20 = 2^2 \times 5$	$8 = 2^3$	9	$10751 = 13 \times$	$1835009 = 11 \times$
1	**41**	**29**	10	**21503**	**7340033**
2	**83**	**113**	11	$43007 = 29 \times$	$29360129 = 37 \times$
3	**167**	**449**	12	$86015 = 5 \times$	$117440513 = 3907 \times$
4	$335 = 5 \times$	$1793 = 11 \times$	13	**172031**	**469762049**
5	$671 = 11 \times$	$7169 = 67 \times$	14	$344063 = 17 \times$	$1879048193 = 11 \times$
6	$1343 = 17 \times$	$28673 = 53 \times$	15	$688127 = 11^4 \times$	$7516192769 = 29^2 \times$
7	**2687**	**114689**	16	$1376255 = 5 \times$	$30064771073 = 113 \times$
8	$5375 = 5^3 \times$	$458753 = 79 \times$	17	$2752511 = 19 \times$	$120259084289 = 379 \times$

Example 24. Consider the sequence

$$x_0 = 1, \quad x_{n+1} = \left(1 + x_0^2 + x_1^2 + \cdots + x_n^2\right)/(n+1) \quad (n \geq 0).$$

n	0	1	2	3	4	5	6	7	8	9	$\cdots$
x_n	1	2	3	5	10	28	154	3520	1551880	267593772160	$\cdots$

Is x_n always an integer?

Example 25. The same, but with cubes in place of squares:

$$y_0 = 1, \quad y_{n+1} = \left(1 + y_0^3 + y_1^3 + \cdots + y_n^3\right)/(n+1).$$

Same question.

n	0	1	2	3	4	5	...
y_n	1	2	5	45	22815	2375152056927	...

Example 26. Also for fourth powers, $z_{n+1} = \left(1 + z_0^4 + z_1^4 + \cdots + z_n^4\right)/(n+1)$.

n	0	1	2	3	4	...
z_n	1	2	9	2193	5782218987645	...

And for fifth powers, and so on.

Example 27. The irreducible factors of $x^n - 1$ are cyclotomic polynomials, i.e., $x^n - 1 = \prod_{d|n} \Phi_d(x)$, so that $\Phi_1(x) = x - 1$, $\Phi_2(x) = x + 1$, $\Phi_3(x) = x^2 + x + 1$, $\Phi_4(x) = x^2 + 1$. The cyclotomic polynomial of order n, $\Phi_n(x)$ has degree $\varphi(n)$, Euler's totient function. It is easy to write down $\Phi_n(x)$ if n is prime, twice a prime, or a power of a prime, and for many other cases. Are the coefficients always ± 1 or 0?

Example 28. If two people play Beans-Don't-Talk, the typical position is a whole number, n, and there are just two options, from n to $(3n \pm 1)/2*$, where $2*$ means the highest power of 2 that divides the numerator. The winner is the player who moves to 1. For example, 7 is a *P-position*, a previous-player-winning position, because the opponent must go to
$$(3 \times 7 + 1)/2 = 11 \quad \text{or} \quad (3 \times 7 - 1)/2^2 = 5$$
and 11 and 5 are *N-positions*, next-player-winning positions, since they have the options $(3 \times 11 - 1)/25 = 1$ and $(3 \times 5 + 1)/2^4 = 1$.

If τ is the probability that a number is an *N*-position, and there are no *O*-positions (from which neither player can force a win), then the probability that a number is a *P*-position is $1 - \tau$. This happens just if both options are *N*-positions, so $1 - \tau = \tau^2$, and τ is the golden ratio, $\left(\sqrt{5} - 1\right)/2 \approx 0.618$.

So it is no surprise that 5 out of the first 8 numbers are *N*-positions, 8 out of the first 13, 13 of the first 21, 21 of the first 34, and 34 of the first 55, since the ratio of consecutive Fibonacci numbers tends to the golden ratio.

Example 29. Does each of the two diophantine equations
$$2x^2(x^2 - 1) = 3(y^2 - 1) \quad \text{and} \quad x(x-1)/2 = 2^n - 1$$
have just the five positive solutions $x = 1, 2, 3, 6$, and 91?

Example 30. Consider the sequence $a_1 = 1$, $a_{n+1} = \left\lfloor \sqrt{2a_n(a_n + 1)} \right\rfloor$ $(n \geq 1)$:

n	1	2	3	4	5	6	7	8	9	10	11	12	13	14	15	16	17	18	19	20	21
a_n	1	2	3	4	6	9	13	19	27	38	54	77	109	154	218	309	437	618	874	1236	1748
		1		2		4		8		16		32		64		128		256		512	

Are alternate differences, $a_{2k+1} - a_{2k}$, the powers of two, 2^k?

Example 31. In the same sequence, are the even ranked members, a_{2k+2}, given by $2a_{2k} + \varepsilon_k$, where ε_k is the kth digit in the binary expansion of $\sqrt{2} = 1.01101010000010...$?

Example 32. Is this the same sequence as $a_1 = 1$, $a_2 = 2$, $a_3 = 3$, $a_{n+1} = a_n + a_{n-2}$ $(n \geq 3)$?

Example 33. The nth derivative of x^x, evaluated at $x = 1$, is an integer. Is it always a multiple of n? Values for $n = 1, 2, 3,...$ are

$$1 \times 1, \ 2 \times 1, \ 3 \times 1, \ 4 \times 2, \ 5 \times 2, \ 6 \times 9, \ 7 \times (-6), \ 8 \times 118, \ 9 \times (-568),$$
$$10 \times 4716, \ 11 \times (-38160), \ 12 \times 358126, \ 13 \times (-3662088),$$
$$14 \times 41073096, \ 15 \times (-500013528), \ 16 \times 6573808200,$$
$$17 \times (-92840971200), \ 18 \times 1402148010528$$

Example 34. In how many ways, c_n can you arrange n pennies in rows, where every penny in a row above the first must touch two adjacent pennies in the row below?

n	0	1	2	3	4	5	6	7	8	9	10	11	12	13	14	15	16
c_n	1	1	1	2	3	5	9	15	26	45	78	135	234	406	704	1222	2120

To throw more light on such sequences, partition theorists often express their generating function

$$\sum_{n=0}^{\infty} c_n x^n = 1 + x + x^2 + 2x^3 + 3x^4 + 5x^5 + 9x^6 + 15x^7 + \cdots$$

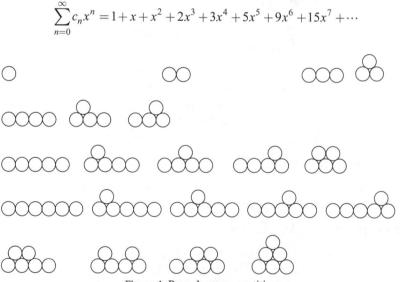

Figure 4. Propp's penny partitions.

as an infinite product,

$$\prod_{n=1}^{\infty}(1-x^n)^{-a(n)}.$$

In this case, $a(n)$ are consecutive Fibonacci numbers:

n	1	2	3	4	5	6	7	8	9	10	...
$a(n)$	1	0	1	1	2	3	5	8	13	21	...

Example 35. If p_k is the kth prime, $p_1 = 2, p_2 = 3,...$, does

$$\prod_{k=1}^{\infty}(1-p_k)^{-1} = 1 + \sum_{k=1}^{\infty}\frac{x^{p_1+p_2+\cdots+p_k}}{(1-x)(1-x^2)\cdots(1-x^k)} ?$$

Answers

1. No less a person than Fermat was fooled by the Strong Law! Euler gave the factorization $2^{32} + 1 = 641 \cdot 6700417$. All other known examples of Fermat numbers are composite; Jeff Young and Duncan Buell [**32**] have recently shown that $2^{2^{20}} + 1$ is composite.

2. There are very few *Mersenne primes*, $2^p - 1$. No one can prove that there are infinitely many; $2^{11} - 1 = 23 \times 89$ is not one. See A3 in [**12**] and sequence 1080 in [**28**].

3. In the "prime number race," $4k - 1$ and $4k + 1$ alternately take the lead infinitely often. This was proved by Littlewood [**18**]. For many papers on this subject see *N*-12 of *Reviews in Number Theory*, for example, Chen [**4**].

4. A theorem of Legendre (see [**6**], for example) states that if D_+ and D_- are the numbers of divisors of n of shapes $4k + 1$ and $4k - 1$, then the number of representations of n as the sum of two squares is $4(D_+ - D_-)$. So $D_+ \geq D_-$ for every number!

5. Before we reveal all, here is a circle (Fig. 5) with ten points to further confuse you. It has 256 regions.

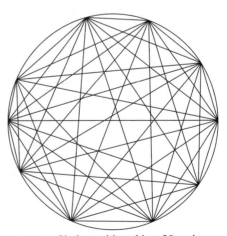

Figure 5. Circle partitioned into 28 regions.

If the circle has n points, there are $\binom{n}{4}$ intersections of chords inside the circle, since each set of four points gives just one such intersection. The number of vertices in the figure is $V = n + \binom{n}{4}$. To find the number of edges, count their ends. There are $n + 1$ at each of the n points and four at each of the $\binom{n}{4}$ intersections, so $2E = n(n + 1) + 4\binom{n}{4}$. By Euler's formula, the number of regions inside the circle is

$$E + 1 - V = 2\binom{n}{4} + \frac{1}{2}n(n+1) + 1 - \binom{n}{4} - n$$

$$= \binom{n}{4} + \frac{1}{2}n(n-1) + 1$$

$$= \binom{n-1}{4} + \binom{n-1}{3} + \binom{n-1}{2} + \binom{n-1}{1} + \binom{n-1}{0}.$$

A direct proof, by labeling the regions with at most four of the numbers 1, 2,..., $n - 1$, will appear in [5]. The answer is just five of the n terms in the binomial expansion of $(1 + 1)^{n-1}$. For $n < 6$, this is all the terms, and the number is a power of 2. For $n = 6$, only 1 is missing. For $n = 10$ just half the terms are missing, and the number of regions is $(1/2)2^9 = 256$.

# of points =	1	2	3	4	5	6	7	8	9	10	11	12	13	14
# of regions =	1	2	4	8	16	31	57	99	163	**256**	386	562	794	1093

Some other famous numbers, e.g. 163 and 1093, also occur in this sequence, number 427 in [28].

6. No member of this sequence is divisible by 2, 3, 5, 7,11,13, or 37, as may be seen immediately from well known divisibility tests. On the other hand, 17, 19, 23, 29, 31,... divide 33...331 just if the number of threes is respectively $16k + 8$, $18k + 11$, $22k + 20$, $28k + 19$, $15k + 1$,..., while 41, 43, 53, 67, 71, 73, 79,... divide no members of the sequence. I don't think that there is a simple description of which primes do, and which primes don't, divide. The next member, 33333331, is also prime, but $333333331 = 17 \times 19607843$.

7. We've again given ourselves a good start, since $\sum_{k=1}^{n}(-1)^{n-k}k!$ is not divisible by any prime $\leq n$. However,

$$9! - 8! + 7! - 6! + 5! - 4! + 3! - 2! + 1! = 326981 = 79 \times 4139.$$

8. This example, as well as example **5**, was first shown to me by Leo Moser, a quarter of a century ago. Row n is the list of denominators of the *Farey series* of order n, i.e., the set of rational fractions r, $0 \leq r \leq 1$, whose denominators do not exceed n. In getting row n from row $n-1$, just $\varphi(n)$ numbers are inserted, where $\varphi(n)$ is Euler's totient function, the number of numbers not exceeding n which are prime to n. It is fortuitous that $1 + \sum_{k=1}^{n} \varphi(k)$ is prime for $1 \le n \le 9$. As $\varphi(10) = 4$, the number of numbers in row 10 is $29 + 4 = 33$, and is not prime.

9. The expression $7013 \times 2^n + 1$ is composite for $0 \le n \le 24160$ [**15**]. Duncan Buell and Jeff Young have sieved out 325 further candidates $n < 10^5$ which might yield a prime. None is known, though it's likely that there is one.

10. The number $78557 \times 2^n + 1$ is always divisible by at least one of 3, 5, 7, 13, 19, 37, 73 [**26**, **27**]. For this and the previous example, see also B21 in [**12**].

11. R. F. Fortune conjectured that these differences are always prime: see [**8**], [**9**] and A2 in [**12**]. The next few are 37, 61, 67, 61, 71, 47, 107, 59, 61, 109, 89, 103, 79. There's a high probability that the conjecture is true, because the difference can't be divisible by any of the first k primes, so the smallest composite candidate for $P = \prod p_k$ is p_{k+1}^2 which is approximately $(k \ln k)^2$ in size. The product of the first k primes is about e^k: to find a counterexample we need a gap in the primes near N of size at least $(\ln N \ln \ln N)^2$. Such gaps are believed not to exist, but it's beyond our present means to prove this.

12. This is N. L. Gilbreath's conjecture, which has been verified for $k < 63419$ [**16**]. Hallard Croft has suggested that it has nothing to do with primes as such, but will be true for any sequence consisting of 2 and odd numbers, which doesn't increase too fast, or have too large gaps: A10 in [**12**]. In an 87-08-03 letter, Andy Odlyzko reported that he had verified the conjecture for $k < 10^{10}$.

13. D. H. and Emma Lehmer discovered that $2^n \equiv 3 \pmod{n}$ for $n = 4700063497$, but for no smaller $n > 1$.

14. The kth Lucas number and the $(k+1)$th Fibonacci number are

$$\left(\frac{1+\sqrt{5}}{2}\right)^k + \left(\frac{1-\sqrt{5}}{2}\right)^k \quad \text{and} \quad \frac{1}{\sqrt{5}}\left\{\left(\frac{1+\sqrt{5}}{2}\right)^{k+1} - \left(\frac{1-\sqrt{5}}{2}\right)^{k+1}\right\}.$$

Their ratio, as k gets large, approaches $\left(5 - \sqrt{5}\right)/2 \approx 1.38196011$, whereas $5^{1/5} \approx 1.379729661$. The next few convergents to $5^{1/5}$,

$$\frac{40}{29}, \frac{109}{79}, \frac{912}{661}, \frac{1021}{740}, \frac{26437}{19161}, \frac{27458}{19901},$$

do not involve Fibonacci or Lucas numbers. Compare sequences 256 & 260 and 924 & 925 in [**28**]. This example goes back to 1866 [**25**].

15. This is quite fortuitous [**30**]. Put $x = y = 1$, giving $2^{2n+1} - 2 = (2n + 1) \times 2 \times 3^{n-1}$. It's true that

$$2^2 - 1 = 3 \times 3^0, \quad 2^4 - 1 = 5 \times 3^1, \quad 2^6 - 1 = 7 \times 3^2$$

but it's clear that the pattern can't continue.

16. The $(n + 1)$th hex number, $1 + 6 + 12 + \cdots + 6n = 3n^2 + 3n + 1$, when added to n^3, gives $(n + 1)^3$, so the pattern is genuine. It is instructive to regard the nth hex number as comprising the three faces at one corner of a cubic stack of n^3 unit cubes (Fig. 6).

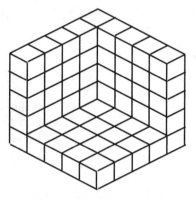

Figure 6. The fifth hex number.

17, 18, 19, and **20** are examples of *Moessner's process*, which does indeed produce the square, cubes, fourth powers and factorials. Moessner's paper [**20**] is followed by a proof by Perron. Subsequent generalizations are due to Paasche [**22**]: see [**19**] for a more recent exposition.

21. A thinly disguised arrangement of Euler's formula, $n^2 + n + 41$, which gives primes for $-40 \leq n \leq 39$. For $n = 40$, $n^2 + n + 41 = 41^2$. See A1 and Fig. 1 in [**12**]. For remarkable connections with quadratic fields, continued fractions, modular functions and class numbers, see [**29**].

22. The initial pattern is explained by the facts that if n is odd, $n^4 + 1$ is even, and $17 \times 2^n - 1$ is a multiple of 3. Thereafter it's largely coincidence until $n = 24$, for which $n^4 + 1 = 331777$ is prime, while $17 \times 2^n - 1 = 285212671 = 149 \times 1914179$. See [**17**], [**24**] and sequences 386 and 387 in [**28**].

23. This is also a coincidence, until we reach $n = 18$, for which $21 \times 2^n - 1 = 5505023$ is prime, while

$$7 \times 4^4 + 1 = 481036337153 = 166609 \times 2887217.$$

See [**31**], [**23**] and sequences 314 & 315 in [**28**].

24. A sequence introduced by Fritz Göbel. A more convenient recursion for calculation is

$$(n+1)x_{n+1} = x_n(x_n + n), \quad (n \geq 1).$$

If you work modulo 43, you'll find that for

$n =$	0	1	2	3	4	5	6	7	8	9	10	11	12	13	14	15	16	17	18	19	20	21
$x_n =$	1	2	3	5	10	28	25	37	10	20	15	38	19	42	36	34	2	35	39	31	13	2

$n =$	22	23	24	25	26	27	28	29	30	31	32	33	34	35	36	37	38	39	40	41	42
$x_n =$	6	26	28	29	4	14	42	5	20	17	4	20	16	29	42	13	42	20	8	23	33

and $x_{42}(x_{42} + 42) \equiv -10(-10 + 42) = -320$, which is not divisible by 43, so x_{43} is not an integer, although x_n is an integer for $0 \leq n \leq 42$.

25. Similar calculations, mod 89, using the relation $(n+1)y_{n+1} + 1 = y_n(y_n + n)$, show that y_{89} is not an integer. For this, and the previous example, see E15 is [**12**].

26. Since this question was asked, Henry Ibstedt has made extensive calculations, and found the first noninteger term, x_n, in the sequence involving kth powers, to be

k	2	3	4	5	6	7	8	9	10	11
n	43	89	97	214	19	239	37	79	83	239

He also found corresponding results with different initial values. The longest to hold out ($n = 610$) are the cubes ($k = 3$, Example **25**) with $x_0 = 1$, $x_1 = 11$.

27. The first cyclotomic polynomial to display a coefficient other than -1 and 0 is

$$\begin{aligned}
\Phi_{105} = {}& x^{48} + x^{47} + x^{46} - x^{43} - x^{42} - 2x^{41} - x^{40} - x^{39} + x^{36} + x^{35} + x^{34} \\
& + x^{33} + x^{32} + x^{31} - x^{28} - x^{24} - x^{22} - x^{20} + x^{17} + x^{16} + x^{15} + x^{14} \\
& + x^{13} + x^{12} - x^9 - x^8 - 2x^7 - x^6 - x^5 + x^2 + x + 1.
\end{aligned}$$

Coefficients can be unboundedly large, but require n to contain a large number of distinct odd prime factors; see [**8**]. More recently, Montgomery and Vaughan [**33**] have shown that if $\Phi_n \sum a(m,n)x^m$ and $L(n) = \ln \max_n |a(m,n)|$ then, for m large,

$$\frac{m^{1/2}}{(\ln 2m)^{1/4}} \ll L(n) \ll \frac{m^{1/2}}{(\ln m)^{1/4}}.$$

28. This game was misremembered by John Conway from John Isbell's game of Beanstalk [**13**]. The Fibonacci pattern is not maintained: only 52 of the first 89 numbers, 81 of the first 144, 126 of the first 233, and 201 of the first 377, are N-positions. The probability argument is fallacious: the probabilities of the status of the two options are not independent.

29. True, but why the coincidence?

30 and **31**. The patterns of powers of 2 and of binary digits of $\sqrt{2}$ both continue; see [**11**], [**14**] and sequence 206 in [**28**].

32. A different sequence, number 207 in [**28**], which agrees for $n < 9$, but then continues 28, 41, 60, 88, 129, 189, 277, 406, 595, 872, 1278,....

33. If $y = x^x$ and $y_n(1)$ denotes the value of $d^n y/dx^n$ at $x = 1$, then

$$y_{n+1}(1) = y_n(1) + \binom{n}{1} y_{n-1}(1) \binom{n}{2} y_{n-2}(1) + 2! \binom{n}{3} y_{n-3}(1) - 3! \binom{n}{4} y_{n-4}(n)$$
$$+ \cdots + (-1)^n (n-1)!.$$

This was not known to be a multiple of $n + 1$ when it was submitted to the Unsolved Problems section of this *Monthly* by Richard Patterson and Gaurar Suri. But in an 87-05-28 letter, Herb Wilf gives a proof, using the generating function for Stirling numbers of the first kind. His proof in fact shows that $n(n - 1)$ divides $y_n(1)$ just if $n - 1$ divides $(n - 2)!$, which it does for $n \geq 7$, provided that $n - 1$ is not prime.

34. This sequence was investigated by Jim Propp. Except that $a(12) = 55$, the pattern of Fibonacci numbers does not continue:

$n =$	11	12	13	14	15	16	17	18
$a(n) =$	35	55	93	149	248	403	670	1082

Since this was written, Wilf [**21**] has linked the generating function with Ramanujan's continued fraction, and he observes that the numbers of propper partitions with k coins in the lowest row are yet another manifestation of the Catalan numbers, 1, 2, 5, 14, 42,... [**7**]. These partitions are a variant of some considered by Auluck [**1**]. Auluck's partitions have the pennies contiguous in every row, not just the lowest. Their numbers 1, 1, 2, 3, 5, 8,... are another good example of the Strong Law.

35. The expansion of the product as a power series is

$$1 + x^2 + x^3 + x^4 + 2x^5 + 2x^6 + 3x^7 + 3x^8 + 4x^9 + 5x^{10} + 6x^{11} + 7x^{12} + 9x^{13}$$
$$+ 10x^{14} + 12x^{15} + 14x^{16} + 17x^{17} + 19x^{18} + 23x^{19} + 26x^{20} + 30x^{21}$$
$$+ 35x^{22} + 40x^{23} + 46x^{24} + 52x^{25} + 60x^{26} + 67x^{27} + 77x^{28} + 87x^{29} + \dots.$$

The sum is the same, until ...
$$+ 31x^{21} + 35x^{22} + 41x^{23} + 46x^{24} + 54x^{25} + 60x^{26} + 69x^{27} + 78x^{28} + 89x^{29} + \dots.$$

This was entry 29 in Chapter 5 of Ramanujan's second notebook [**2**], [**3**]: but he had crossed it out!

Let me know if I've missed your favorite example!

References

1. F. C. Auluck, On some new types of partitions associated with generalized Ferrers graphs, *Proc. Cambridge Philos. Soc.*, 47 (1951) 679–686; MR 13, 536.

2. Bruce C. Berndt, *Ramanujan's Notebooks*, Part 1, Springer-Verlag, 1985, p. 130.

3. Bruce C. Berndt and B. M. Wilson, Chapter 5 of Ramanujan's second notebook, in M. I. Knopp (ed.) *Analytic Number Theory*, Lecture Notes in Math. 899, Springer, 1981, pp. 49–78; MR 83i:10011.

4. W. W. L. Chen, On the error term of the prime number theorem and the difference between the number of primes in the residue classes modulo 4, *J. London Math. Soc.*, (2) 23(1981) 24–40; MR 82g:10058.

5. John Conway and Richard Guy, *The Book of Numbers*, Scientific American Library, W. H. Freeman, 1988.

6. H. Davenport, *The Higher Arithmetic*, Hutchinson's University Library, 1952, p. 128.

7. Roger B. Eggleton and Richard K. Guy, Catalan Strikes again! How likely is a function to be convex?, *Math. Mag.* 61(1988) 211–218.

8. P. Erdős and R. C. Vaughan, Bounds for the r-th coefficients of cyclotomic polynomials, *J. London Math. Soc.*, (2) 8(1974) 393–400; MR 50 #9835.

9. Martin Gardner, Mathematical games: patterns in primes are a clue to the strong law of small numbers, *Sci. Amer.*, 243 #6 (Dec. 1980) 18, 20, 24, 26, 28.

10. Solomon W. Golomb, The evidence for Fortune's conjecture, *Math. Mag.* 54(1981) 209–210.

11. R. L. Graham and H. O. Pollak, Note on a linear recurrence related to $\sqrt{2}$, *Math. Mag.*, 43(1970) 143–145; MR 42 #180.

12. Richard K. Guy, *Unsolved Problems in Number Theory*, Springer, 1981.

13. Richard K. Guy, John Isbell's game of Beanstalk and John Conway's game of Beans-Don't-Talk, *Math. Mag.*, 59(1986) 259–269.

14. F. K. Huang and S. Lin, An analysis of Ford and Johnson's sorting algorithm, *Proc. 3rd Annual Princeton Conf. on Info. Systems and Sci.*

15. Wilfrid Keller, Factors of Fermat numbers and large primes of the form $k \times 2^n + 1$, *Math. Comput.*, 41(1983) 661–673.

16. R. B. Killgrove and K. E. Ralston, On a conjecture concerning the primes, *Math. Tables Aids Comput.*, 13(1959) 121–122; MR 21 #4943.

17. M. Lal, Primes of the form $n^3 + 1$, *Math. Comput.*, 21(1967) 245–247.

18. J. E. Littlewood, Sur le distribution des nombres premiers, *C. R. hebd. Séanc. Acad. Sci., Paris*, 158 (1914) 1868–1872.

19. Calvin T. Long, Strike it out—add it up, *Math. Mag.*, 66(1982) 273–277. See also *Amer. Math. Monthly* 73(1966) 846–851.

20. Alfred Moessner, Eine Bemerkung über die Potenzen der natürlichen Zahlen, *S.-B. Math.-Nat. Kl. Bayer. Akad. Wiss.*, 1951, 29(1952); MR 14-353b.

21. Andrew M. Odlyzko and Herbert S. Wilf, *n* coins in a fountain (to appear in *Amer. Math. Monthly.* (Nov. 1988)).

22. Ivan Paasche, Eine Verallgemeinerung des Moessnerschen Satzes, *Compositio Math.*, 12(1956) 263-270; MR 17 836g.

23. Hans Riesel, Lucasian criteria for the primality of $N = h \cdot 2^n - 1$, *Math. Comput.*, 23(1969) 869–875.

24. Raphael M. Robinson, A report on primes of the form $k \cdot 2^n + 1$ and on factors of Fermat numbers, *Proc. Amer. Math. Soc.*, 9(1958) 673–681.

25. P. Seeling, Verwandlung der irrationalen Grösse $\sqrt[n]{\ }$ in einen Kettenbruch, *Archiv. Math. Phys.*, 46(1866) 80–120 (esp. p. 116).

26. J. L. Selfridge, Solution to problem 4995, this *Monthly*, 70(1963) 101.

27. W. Sierpinski, Sur un problème concernant les nombres $k \cdot 2^n + 1$, *Elem. Math.*, 15(1960) 73–74; MR 22 #7983; corrigendum, *ibid.* 17(1962) 85.

28. N. J. A. Sloane, *A Handbook of Integer Sequences*, Academic Press, 1973.

29. Harold M. Stark, An explanation of some exotic continued fractions found by Brillhart, *Computers in Number Theory*, Atlas Sympos. No. 2, Oxford, 1969, pp. 21–35, Academic Press, London, 1971.

30. Peter Taylor and Doug Dillon, Problem 3, *Queen's Math. Communicator*, Dept. of Math. and Statist., Queen's University, Kingston, Ont., Oct. 1985, p. 16.

31. H. C. Williams and C. R. Zarnke, A report on prime numbers of the forms $M = (6a + 1)2^{2m-1} - 1$ and $M' = (6a + 1)2^{2m-1}$, *Math. Comput.*, 22(1968) 420–422.

32. Jeff Young and Duncan A. Buell, The twentieth Fermat number is composite, *Math. Comput.* 50(1988) 261-263.

33. H. L. Montgomery and R. C. Vaughan, The order of magnitude of the *m*th coefficients of cyclotomic polynomials, *Glasgow Math. J.* 27(1985) 143–159; MR 87e:11026.

In 2002, Richard Guy wrote of himself (in the *College Mathematics Journal*, 33 (2002) #3, 188)

Richard Guy is the luckiest person in the world. He's been paid to enjoy mathematics from kindergarten to post-graduate level in Europe, Asia, and America. He's been privileged to work with Erdős, Conway, and Berlekamp.

More than fifty other co-authors include Lehmer, Selfridge, Knuth, Matijasevich, and Martin Gardner. At 85 he's working on becoming the oldest inhabitant of the University of Calgary, where he continues to put in a full day's play. He and Louise [his wife] still hike and ski in the mountains.

He is the author of many books, including *Winning Ways for Your Mathematical Plays* (with Elwyn Berlekamp and John H. Conway; 1990, second edition 2001–2), *The Book of Numbers* (with John H. Conway, 1996), *Unsolved Problems in Number Theory* (1995), and *The Inquisitive Problem Solver* (with Paul Vanderlind and Loren Larson, 2002).

A sequel to his law of small numbers paper is "The Second Strong Law of Small Numbers", *Mathematics Magazine* 63 #1 (1990), pp. 3–20.

9
The Parallel Postulate

Euclid was amazing. This can be said even though we know nothing about him other than that he lived in Alexandria some time during the reign of the Ptolemy who ruled between 325 and 285 BC and was the author of the *Elements*. There is an anecdote or two told about him, but, given the propensity of the human race to make up stories about extraordinary people, there is no reason to think that they ever occurred as related. One, though, is worth repeating. Euclid was asked by a student, the anecdote goes, that perennial question, "What is all this good for?" Euclid instructed his slave to give the student the equivalent of a dollar, so that he could say that he had gained something from mathematics. Modern teachers of mathematics have neither slaves nor enough dollars to follow Euclid's example, even if they wanted to.

Euclid was amazing because the *Elements* was so well done that it lived on, as the first (and until the Renaissance, the last) word on geometry for two thousand years, a longer life than has been had by any other non-religious book. Euclid did not create all the mathematics in the *Elements*, and it is even possible that he did nothing new, but he was a superb compiler and organizer. (There is an anecdote about Blaise Pascal—this shows how far anecdotes can be depended on—that when he was a child he reinvented geometry for himself, independently proving the theorems in the *Elements*, and doing them *in the same order as Euclid*. The *Elements* were so close to holy writ that the deviser of the anecdote must have thought that any departure from Euclid's system would be heresy.)

On the basis of intuition alone, I think that Euclid in fact did not originally prove any of the theorems in the *Elements*. In our day, many mathematicians, if not most of them, are a bit impatient with proof. The fun in doing mathematics is getting the new result and seeing where it leads. Writing it up in a proper form is, if not drudgery, then considerably less fun. Euclid, I think, was a magnificent editor, taking the results of his predecessors, some of them seeming to him muddled, and by using his wonderfully clear mind, putting them in the proper form, shape, and order. This is the merest speculation, but one of the purposes of history is to give opportunities for such things.

The basis for Euclidean geometry is Euclid's five postulates. Here they are, paraphrased:

1. A straight line can be drawn joining any two points.
2. Finite straight lines can be extended indefinitely.
3. A circle can be drawn with any center and radius.
4. All right angles are equal.
5. If a straight line falling on two straight lines makes the interior angles on the same side less than two right angles, then the two straight lines, if extended, will meet on the side where the angles are less than two right angles.

The first three postulates set out the tools of Euclidean geometry. The first says that we may use a straightedge and the second says that it is indefinitely long. The third says that we may use a compass. The fourth tells us that the Euclidean plane is *uniform*, looking and feeling the same no matter where we are on it.

The fifth is different, and not only because it is at least four times as long as the other four. It says (see Fig. 1) that when two lines AB and CD are cut by another, EF, most of the time one of the two angle sums, $= \angle 1 + \angle 2$ or $\angle 3 + \angle 4$, will be less than two right angles. If it is $\angle 1 + \angle 2$, the lines will intersect to the right. If it is $\angle 3 + \angle 4$, they will intersect to the left. If $\angle 1 + \angle 2 = \angle 3 + \angle 4$, the postulate is silent.

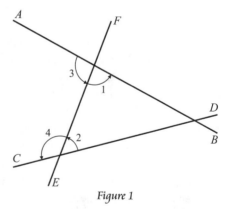

Figure 1

This is obvious enough, though it seems complicated. Why not say, as many geometry books do, that through a point outside a line, one and only one parallel can be drawn through the point? Because—and this is part of Euclid's genius—that would be *more* complicated, involving as it does a new idea, "parallel", that would have to be defined.

If parallel lines are defined to be lines that always have the same distance from each other, then distance would have to be defined. Because distance is not necessary for plane geometry—the Euclidean straightedge has no marks on it and the Pythagorean theorem, $a^2 + b^2 = c^2$, is about areas of squares, not about lengths of lines—it is more economical and elegant not to include it. If parallel lines are defined to be lines that never intersect, then the postulate becomes difficult to check. To assert that lines AB and CD intersect on the sides of the smaller angle sum involves examining only a finite portion of the plane, but to see that lines *never* meet you must look at the whole plane, infinite in extent. This takes very good eyesight. Euclid presented his fifth postulate with care, in the best way.

Even so, to many mathematicians, the fifth postulate seemed to be a defect in the system, an unsightly addition. How much nicer it would be if only the first four postulates were needed! In the excerpt that follows, from *The Non-Euclidean Revolution*, Richard Trudeau presents, among other things, a large and surprising variety of statements that are equivalent to the fifth postulate.

The Problem With Postulate 5

Richard J. Trudeau (1987)

For 2100 years after the appearance of the *Elements*, a steady trickle of subtle thinkers were disturbed by Postulate 5. It wasn't as simple as the other axioms. No one doubted it was true, but it seemed out of place as a basic assumption.

At first the problem was perceived as simply aesthetic. Postulate 5 "sounded" more like a theorem than an axiom. The fact that its converse actually is a theorem (Theorem 17) reinforced the feeling that it need not be assumed, but could be proved. As mathematicians have always considered it "inelegant" to assume more than is absolutely necessary, this led to the conviction that Postulate 5 *should* be proved.

Later, because of a misinterpretation of Euclid's intentions, the problem acquired a philosophic dimension as well. Euclid himself seems not to have considered the truth of his postulates to be obvious. "As regards the postulates," concludes Sir Thomas L. Heath after a lengthy analysis (Heath's *Euclid*, pp. 117–124), "we may imagine [Euclid] saying:

> Besides the common notions there are a few other things which I must
> assume without proof, but which differ from the common notions in that

they are not self-evident. The learner may or may not be disposed to agree to them; but he must accept them at the outset on the superior authority of his teacher, and must be left to convince himself of their truth in the course of the investigation which follows."

But as Euclid never made this explicit, the idea inevitably arose that he had intended not only the common notions, but also the postulates, to be acceptable at the outset as obviously true—a standard, many would feel, the complicated Postulate 5 did not meet. This gave new urgency to the problem with Postulate 5: its status as an axiom was more than mathematically inelegant—it was philosophically objectionable.

A number of historians have inferred from the structure of Book I that Euclid himself was somehow uncomfortable with Postulate 5. The fact that until Theorem 29 no theorem depends on it, and thereafter *every* theorem except Theorem 31 depends on it, makes it appear as if he wanted to postpone using it as long as he could. Further, there are theorems that he takes the trouble to prove without Postulate 5, though he could have proven them much more easily had he waited for Postulate 5 to appear on the scene—like the AAS part of Theorem 26, which is an immediate consequence of Theorem 32.

Poseidonios

The first person we know definitely to have been uncomfortable with Postulate 5 was the philosopher, scientist, and historian Poseidonios[1] (c. 135–c. 51 BC), who suggested proving Postulate 5 by changing Definition 23 to read

Parallel straight lines are straight lines which, being in the same plane and being produced indefinitely in both directions, keep always the same distance between them.

The defining property is called "equidistance" and is certainly well in accord with our mental image of parallel lines. In fact I suspect a layman asked to explain "parallel" lines would sooner give Poseidonios' definition than Euclid's.

You may be puzzled when I speak of "proving Postulate 5." In a given system of geometry one of course does not prove a postulate. Poseidonios' plan was to reorganize Euclidean geometry by striking the statement known to us as "Postulate 5" from the list of axioms and placing it instead among the theorems where it seemed to belong. In Poseidonios's proposed system his "proof of Postulate 5" would be the proof of a theorem. When I refer to "proving Postulate 5" now or in the sequel, I mean "proving the statement that in Euclid's system was called 'Postulate 5'."

Poseidonios changed the definition of "parallel" because a proof of Postulate 5 using Euclid's definition somehow escaped him. How could this be? Don't the two definitions amount to the same thing?

A little reflection on this anomaly reveals that Poseidonios' plan, unless modified, will not work. He clearly *intended* his definition of "parallel" to "amount to the same thing" as Euclid's, that is, to apply in exactly the same situations. For example, if *AB* is a given straight line, he intended that every straight line "parallel" to *AB* in his sense of the word (equidistant from *AB*) would also be "parallel" to *AB* in Euclid's sense (never meeting *AB*), and vice-versa. Though it certainly seems impossible to imagine a straight line parallel to *AB* in one sense but not the other, the issue here is not what is imaginable, but what is logical.

To make things more explicit, Poseidonios' plan stands or falls according as the mathematical community is certain or doubtful of the following two statements:

(1) If two straight lines are equidistant no matter how much they are produced, then they never meet.

(2) If two straight lines never meet no matter how much they are produced, then they are equidistant.

Of course no one has ever doubted (1). If two straight lines keep always the same (positive) distance between them, then the distance between them can never be zero. But what of (2)? Straight lines can be produced indefinitely, so we can never examine them over their entire potential lengths. Even if we know a certain pair of straight lines never meet, no matter how much they are produced, and we find them to be equidistant over the tiny portion of their potential lengths to which we have access, how can we be sure the distance between them doesn't fluctuate somewhere beyond our reach? (If you answer, "Because they are *straight*!", remember that "straight line" is a primitive term and the only properties of primitive terms we can use are those contained in the axioms.) I am not quibbling. In mathematics a request for a supporting argument is *always* legitimate. We need a proof of (2).

We already have one, practically. That parallel lines (in Euclid's sense) are equidistant is an easy consequence of Theorem 34. But such a proof won't do here, as Theorem 34 depends on Postulate 5, the very postulate Poseidonios has struck from the roster!

Actually it is known that, given Poseidonios' foundation, *no* proof of (2) is possible. (2) must be *postulated*, as Poseidonios tacitly did when he assumed his definition of "parallel" to be equivalent to Euclid's. And since postulating

(2) explicitly would make his new definition of "parallel" redundant, we may as well keep the old one.

The result, after the dust settles, is a plan for reorganizing Euclidean geometry that is practicable but less dramatic than Poseidonios' original proposal. In it Postulate 5 is still demoted to Theorem, but its place, which Poseidonios would have left vacant, has now been taken by statement (2), which from now on I will call "Poseidonios' Postulate."

Poseidonios' Reorganization (Modified)

Foundation

Keep Euclid's primitive terms and Definitions of defined terms (including Euclid's Definition of "parallel");

keep Euclid's Common Notions, Postulates 1–4 and Postulates 6–10;

replace Postulate 5 with Poseidonios' Postulate, which (since we are keeping the original Definition of "parallel") now says: "Parallel straight lines are equidistant."

Theorems

Keep the theorems proven without Postulate 5, with their original proofs;

state and prove a new theorem whose content is that of Euclid's Postulate 5;

keep the theorems Euclid proved *with* Postulate 5, with their original proofs, except that citations of "Postulate 5" are changed to citations of the new theorem.

Note that to effect the entire scheme it remains only to prove the "new theorem," i.e., to prove Postulate 5.

[Proof omitted]

Metageometry

Though the proofs of the last section resemble those of Chapter 2, our purpose in constructing them was different. We wanted to establish, not facts about straight lines in a plane, but rather that Euclid's Postulate 5 can be proven as a theorem in Poseidonios' reorganization—a fact about an *entire geometric system*. Modern mathematicians call such facts "*metageometrical*."

In *geometry* we examine geometric figures and report (in "axioms" and "theorems") on what we see. Thus the objects of our study are *figures*, and the system we construct is a list of statements about those figures. In *metageom-*

etry, on the other hand, the objects of study are *geometric systems* themselves, and so our reports on what we see ("metatheorems") are statements about *statements*. For example, "In isosceles triangles the angles at the base are equal" is a statement about geometric figures and is part of Euclid's system it—is, in fact, Theorem 5; but the statement, "Euclid's Theorem 5 can be proven without Postulate 5" is a metatheorem because it does not concern geometric figures (directly), but rather the logical relationship of certain statements *about* geometric figures in Euclid's system.

In this light the proofs of the last article can be seen as the proofs of two metatheorems:

Metatheorem 1. *In Poseidonios' reorganization it is possible to prove that:*

Through a given point not on a given straight line, and not on that straight line produced, no more than one parallel straight line can be drawn.

Metatheorem 2. *In Poseidonios' reorganization it is possible to prove Postulate 5.*

In metageometry the methods are familiar, but the perspective from which we work, and so the import of our work, are new. Renaming our results recognizes these differences.

Evaluation of Poseidonios' Reorganization

Poseidonios wanted to somehow "improve" Euclidean geometry by showing the complicated Postulate 5 to be a logical consequence of Euclid's other axioms. Feeling that his definition of "parallel" was equivalent to Euclid's, he thought he knew how to do this.

Unfortunately, as we have seen, his definition comprises not only Euclid's definition, but also a tacit assumption that Euclid's parallels are equidistant. Carrying out Poseidonios' reorganization, therefore, does not consist in deducing Postulate 5 from the other axioms alone, as he had thought, but in deducing it from the other axioms *plus* his unconscious assumption (Poseidonios' Postulate). This we have done; but how close have we come to achieving Poseidonios' goal? Is Poseidonios' organization of Euclidean geometry any "better" than Euclid's?

At first blush it seems that it is. Reading Poseidonios' Postulate and Euclid's Postulate 5 side by side, the former strikes us as simpler, easier to understand (it "sounds" more like an axiom), and something a novice would more quickly accept (it seems more "self-evident").

Recalling my earlier objection, however, we can see that Euclid's postulate, for all its complexity, may be the one that is more "evident." A straight line can be produced indefinitely, and we have access to only a limited portion of it. Given a pair of straight lines that do not meet, we could not possibly check that they are equidistant over their entire potential lengths, as Poseidonios' Postulate asserts, because we cannot travel (even via telescope) indefinitely far. Conceivably (though admittedly difficult to imagine) the distance between the lines might fluctuate somewhere beyond our reach. As this applies to any pair of straight lines that do not meet, we cannot directly verify a single instance of Poseidonios' Postulate. We can, however, verify some instances of Euclid's postulate. Given two straight lines cut by a third so that the interior angles on the same side add up to less than 180°, Euclid's postulate asserts that the two straight lines will meet at a point that is only a finite distance away; if the angle-sum is not too close to 180° we will be able to travel to that point and see for ourselves.[2] And having verified the postulate whenever the point of intersection is reachable, we will tend to find it plausible even when the angle-sum is almost 180° and the point impracticably far.

Up to now we have been comparing the two postulates in psychological terms. But on mathematical issues the deciding factor is usually logical, so let's see what logicians have to say.

Modern logicians call two statements arising in some context "logically equivalent" if, in that context, each is deducible from the other. That is, if the two statements are A and B, then A and B are *logically equivalent* if

(1) Context $+ A \Rightarrow B$, and

(2) Context $+ B \Rightarrow A$.

(The meaning of the double-shafted arrow—read "implies"—is that the statement it points to can be deduced from those before it.) In this situation logicians use the word "equivalent" (from Latin, "of equal value") because whenever A is true, so is B—by (1)—and whenever B is true, so is A—by (2)—so in the given context they are *equally* true; the truth of one is *related* to the truth of the other, and logicians don't care about anything (professionally!) except how statements are *related*. It doesn't matter if A and B seem to say very different things. Logic is blind to a statement's apparent substance, or tone, or complexity, or intuitive impact, or any other psychological feature. In deduction the only significant question one can ask about a statement is, "In view of what we have, does it follow?" A and B have the same logical value because, in the given context, the answer to that question for A would be the same as the answer for B.

Presently the context is the common part of Poseidonios' and Euclid's foundations—Euclid's primitive terms, Definitions of defined terms, Common Notions, and Postulates other than #5, to which we can add all the theorems that are deducible therefrom. I will call this context "Neutral geometry,[3]" as it is the part of Euclidean geometry that is noncontroversial. The statements whose relative logical value we would like to assess are Poseidonios' Postulate and Euclid's Postulate 5. They will be logically equivalent if

(1) Neutral geometry + Poseidonios' Postulate ⇒ Euclid's Postulate 5, and

(2) Neutral geometry + Euclid's Postulate 5 ⇒ Poseidonios' Postulate.

We already have implication (1), because what it asserts is precisely what occupied us in the last section ((1) is Theorem P2). We practically have implication (2) as well. We said on p. 120 that the equidistance of parallel straight lines in Euclid's system is an easy consequence of Theorem 34, so verifying implication (2) is just a matter of writing down those few steps, which I leave to you. Thus we have

Metatheorem 3. *In the context of Neutral geometry, Poseidonios' Postulate and Euclid's Postulate 5 are logically equivalent.*

Had implication (2) turned out to be false, Poseidonios' Postulate would have been logically prior, more "basic" (and more "sweeping") than Euclid's. It would have entailed Euclid's but Euclid's would not have entailed it. In that case substituting Poseidonios' Postulate for Euclid's would have had considerable value by simplifying the logical structure of Euclidean geometry's foundation, but as things have turned out, a logician would say there is simply no point to adopting Poseidonios' reorganization.

We have considered three viewpoints, each coming to its own conclusion on the relative merit of Poseidonios' and Euclid's systems. To summarize: the first view—which without prejudice I will call the "naive"—view is that Poseidonios' Postulate is better because it is shorter and easier to comprehend (more as an axiom "should" be); the second—I will call it the "scientific" view—is that Euclid's postulate is better because it is experimentally verifiable at least in part; and the third or "logical" view is that neither is preferable because they are logically equivalent.

The choice is not an easy one. Behind the various views are different opinions—all well- established—as to the very nature of Euclid's work. Is the *Elements* (a) a textbook, or (b) a work of art? Is it (c) a scientific description of space? Or (d) an exercise in pure logic? From the beginning all four aspects have been present, with different emphases, in the ways scholars have perceived the book.

It's easy to see how believing the *Elements* to be primarily (a) or (b) leads to the naive viewpoint. For then the difficulty with Postulate 5 is either pedagogical (it is difficult for a beginner) or aesthetic (it is a blemish on Euclid's beautiful system), but in any case bound up with Postulate 5's *unwieldiness*. Poseidonios' Postulate is then an improvement. Similarly finding the *Elements* to be primarily (c) causes one to interpret the controversy as concerning the *reliability of a scientific principle* and so tilts the balance in favor of a postulate subject to experimental test. This is the scientific view, and favors Euclid's postulate. Finally, believing the *Elements* to be primarily (d) translates the discomfort with Postulate 5 into a longing for a logically simpler foundation for geometry. This is the logical view, according to which replacing Postulate 5 with a logically equivalent statement (such as Poseidonios' Postulate) is pointless.

Let me caution you that I'm making the alternatives seem fewer and more clear-cut than they really are. The naive, scientific and logical views and the corresponding perceptions of the *Elements* are only *general postures* I've proposed to call attention to the various issues that arise when we contemplate substituting another postulate (and not necessarily Poseidonios') for Postulate 5. The actual viewpoints of living, thinking people can't always be placed neatly in categories. A person might, for example, consider the *Elements* to be primarily a treatise on physical space ("scientific view"), but still favor Poseidonios' Postulate on grounds that in my categorization are associated with the other perspectives. His/her reasoning might go like this:

A. Principles should be as simple as possible ("naive view").

B. Since Poseidonios' Postulate is logically equivalent to Euclid's postulate ("logical view"), an experiment tending to confirm one simultaneously tends to confirm the other.

C. Thus the experimental advantage of Euclid's postulate is illusory.

D. Therefore, other things being equal, Poseidonios' Postulate is better because it is simpler.

And so on. You may yourself have forged a chain of reasoning, in support of one postulate or the other, that is different still.

The only way to settle the question is by fiat. I will, from now on in this chapter, concern myself chiefly with the posture I have called the "logical view." I will do this not because I know of any crippling objections that can be made to the other viewpoints, nor because I have been holding in reserve some overwhelming argument in its favor, but simply because it is the viewpoint that *kept the controversy alive* and thereby led, ultimately, to non-Euclidean geometry.

Overview of Later Attempts

Between the first century BC and the early 19th century scores of thinkers were disturbed by Postulate 5 and set about trying to prove it. Those who succeeded invariably made use of an extra assumption. Consequently all the many proposals boil down to the same general plan we uncovered in our analysis of Poseidonios' proposal:

(1) Replace Postulate 5 with a more acceptable assumption;

(2) leave the rest of Euclid's foundation (i.e., the foundation of Neutral geometry) intact;

(3) prove Postulate 5.

Often, especially at first, the replacement postulate was substituted unconsciously, and for a while people thought (as had happened to Poseidonios) that Postulate 5 had been deduced from the Neutral foundation *alone*. This would have meant that the troublesome Postulate 5 could simply be removed from Euclidean geometry's foundation, that what remained would be able to support the entire superstructure of theorems unassisted, and thus that Euclidean geometry and Neutral geometry were the same. But in every case the hidden assumption was eventually ferreted out by some later thinker.

Other times the mathematician was well aware of his replacement postulate, but felt he had found one with which the mathematical world could at last be content. The English mathematician John Wallis (1616-1703), for example, proposed replacing Postulate 5 with the following.

Wallis' Postulate. On a given finite straight line it is always possible to construct a triangle similar to a given triangle.[4]

("Similar" triangles have the same three angles, and so the same shape.) You may find it difficult to imagine how Wallis could have deduced Postulate 5 from this, as it seems to say a very different thing; but logically it is not much further from Postulate 5 than Poseidonios' Postulate was. Wallis defended his postulate on the ground that what it asserts for triangles is very nearly what Postulate 3 already does for circles, i.e., the existence of figures having the same shape but arbitrary size.

A few mathematicians, especially during the 18th and early 19th centuries, were less easily satisfied.

> As for me, I have already made some progress in my work. However the path I have chosen does not lead at all to the goal which we seek [proving Postulate 5].... It seems rather to compel me to doubt the truth of geometry itself.

It is true that I have come upon much which by most people would be held to constitute a proof: but in my eyes it proves as good as *nothing*. For example, if one could show that a ... triangle is possible, whose area would be greater than any given area, then I would be ready to prove the whole of geometry absolutely rigorously.

Most people would certainly let this stand as an Axiom; but I, no! It would, indeed, be possible that the area might always remain below a certain limit, however far apart the three angular points of the triangle were taken.

— letter[5] from Carl Friedrich Gauss
to Farkas Bolyai, December 17, 1799

(People who like to compare mathematicians frequently rank the German mathematician, physicist, and astronomer Gauss (1777–1855) with Archimedes and Isaac Newton in a three-way tie for all-time greatest. You will see Gauss' name again: a few years after writing this letter, he became one of the inventors of non-Euclidean geometry. Gauss' lifelong friend Farkas Bolyai (1775–1856) was a Hungarian mathematician whose son, János Bolyai (1802–1860), became another of the inventors of non-Euclidean geometry.) Gauss' replacement postulate I will naturally call

Gauss' Postulate. It is possible to construct a triangle whose area is greater than any given area.

Psychologically this is even further from Postulate 5 than Wallis' Postulate was, and now the logical distance is great as well; nonetheless Gauss did manage, amazingly, to deduce Postulate 5. But note his displeasure at having to resort to *any* replacement postulate *at all*.

Here are some of the more noteworthy replacement postulates that have been proposed over the years, grouped by surface similarity; most are followed by the names of the mathematicians who proposed them.

1A. Parallel straight lines are equidistant. (Poseidonios, 1st century B.C.)

1B. All the points equidistant from a given straight line, on a given side of it, constitute a straight line. (Christoph Clavius, 1574)

1C. In any quadrilateral having two equal sides perpendicular to a third side (Figure 104) there is at least one point (H in the figure) on the fourth side from

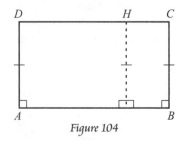

Figure 104

which the perpendicular to the side opposite is equal to the two equal sides. (Giordano Vitale, 1680)

1D. There exists at least one pair of equidistant straight lines.

2. The distance between a pair of parallel infinite straight lines (may fluctuate, but) remains less than a certain fixed distance. (Proclus, 5th century)

3A. (Theorem 30) Straight lines parallel to the same straight line are also parallel to one another.

3B. If a straight line intersects one of two parallel straight lines, it will, if produced sufficiently, intersect the other (or the other produced) also.

3C. Through a given point not on a given straight line, and not on that straight line produced, no more than one parallel straight line can be drawn. (popularized by John Playfair, late 18th century)

4A. If two straight lines (*AB*, *CD* in Figure 105) are cut by a third (*PQ*) that is perpendicular to only one of them (*CD*), then the perpendiculars from *AB* to *CD* are less than *PQ* on the side on which *AB* makes an acute angle with *PQ*, and are greater on the side on which *AB* makes an obtuse angle. (Nasir al-Din, 13th century)

4B. Straight lines which are not equidistant converge in one direction and diverge in the other. (Pietro Antonio Cataldi, 1603)

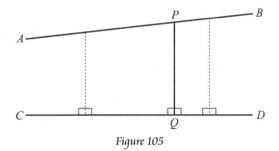

Figure 105

5A. On a given finite straight line it is always possible to construct a tri angle similar to a given triangle. (John Wallis, 1663; Lazare-Nicholas; Marguerite Carnot, 1803; Adrien-Marie Legendre, 1824)

5B. A pair of noncongruent, similar triangles exists. (Gerolamo Saccheri, 1733)

6A. In any quadrilateral having two equal sides perpendicular to a third side (Figure 104), the other two angles are right. (Gerolamo Saccheri, 1733)

6B. In any quadrilateral having three right angles, the fourth angle is also right. (Alexis-Claude Clairaut, 1741; Johann Heinrich Lambert, 1766)

6C. At least one rectangle exists. (Gerolamo Saccheri, 1733)

7A. (Theorem 32 (b)) The angle-sum of every triangle is 180°. (Gerolamo Saccheri, 1733; Adrien-Marie Legendre, early 19th century)

7B. At least one triangle with angle-sum 180° exists. (same)

8.   There is no absolute standard of length. (Johann Heinrich Lambert, 1766; Adrien-Marie Legendre, early 19th century)

9A. Every straight line through a point within an angle will, if produced sufficiently, meet at least one side of the angle (or one side produced). (J. F. Lorenz, 1791)

9B. Through any point within an angle of less than 60°, it is always possible to draw a straight line which meets both sides (produced if necessary) of the angle. (Adrien-Marie Legendre, early 19th century)

10. It is possible to construct a triangle whose area is greater than any given area. (Carl Friedrich Gauss, 1799)

11. In the plane translations and rotations of a straight line are independent motions. (Bernhard Friedrich Thibaut, 1809)

12. Given three points not on a single straight line, it is always possible to draw a circle passing through all three. (Adrien-Marie Legendre, Farkas Bolyai, early 19th century)

Over the years holders of the naive view, for whom the chief difficulty with Postulate 5 was its unwieldiness, were relatively quickly satisfied, if not by Poseidonios' Postulate—which they may not have heard of—then by some other concise alternative (for instance "Playfair's Postulate" 3C, which has been widely used in textbooks). And substitutes for Postulate 5 were discovered that were completely testable by experiment (for instance the Saccheri-Legendre postulate 7B, which involves only measuring the angles of a *single* triangle), so holders of the scientific view were eventually satisfied as well. But holders of the logical view were frustrated time and again.

Let's return to Gauss, who considered his very own postulate "as good as nothing" (p. 127), even though what it asserts for areas of triangles that they can be made arbitrarily large is no more than what Postulate 2 asserts for lengths.

Gauss realized, I think, that his postulate is logically equivalent to Postulate 5. Remember that a statement "A" is logically equivalent to Postulate 5 if

(1) Neutral geometry $+ A \Rightarrow$ Postulate 5, and

(2) Neutral geometry + Postulate 5 $\Rightarrow A$,

where "Neutral geometry" consists of Euclid's primitive terms, Definitions of defined terms, Common Notions, Postulates other than #5, and all theorems derivable therefrom. Taking "*A*" to be Gauss' Postulate, Gauss himself had established (1) when he proved Postulate 5. And, although we didn't go that far in Chapter 2, it is well-known that in Euclid's geometry one can construct a triangle with area as large as one pleases, which (since Neutral geometry + Postulate 5 = Euclid's geometry) is what (2) says.

You may wonder why Gauss, if he realized the logical equivalence, didn't say so in his letter to Farkas Bolyai. He didn't because he didn't have the vocabulary. In mathematics it takes a long time for a new abstract concept to emerge and then come into focus, and until this happens it can't be sharply defined. "Logical equivalence of geometric postulates" is a notion not formulated precisely until long after Gauss wrote his letter, at a time when mathematicians were shocked by non-Euclidean geometry into looking at all geometry more abstractly than ever before. When I attribute the "logical view" to Gauss and other mathematicians of the early 19th century and before (especially Saccheri, Lambert, Legendre, and Farkas Bolyai), I am interpreting their work in modern terms they would never have used. Nonetheless their espousal of the logical view is apparent in their continual dissatisfaction with replacement postulates, and their dogged striving to prove Postulate 5 without resorting to one.

This persistent unwillingness of Gauss and his intellectual brothers to accept a replacement postulate indicates that they were aware, though we can only guess how clearly, of a surprising thing I haven't mentioned yet, although it may have already occurred to you: that *any* replacement postulate (any of the ones listed on pp. 128–129, for example) is logically equivalent to Postulate 5! Nowadays this is easy to see, in fact by practically the same argument we used two paragraphs ago. For whoever proposes a replacement postulate "*A*" will of course use it to prove Postulate 5—that's the whole point—and thereby establish (1). But the proposed postulate *A* will surely be something that is provable in Euclidean geometry (as originally organized), and thus (2) will occur as well! Thus no replacement postulate can be logically prior to Postulate 5; they are *all* logically equivalent to it and therefore, in the logical view, all worthless.

Suddenly the frustration of Gauss and the other logical-view mathematicians is understandable. They could see no way of proving Postulate 5 without a replacement postulate. Yet—and in varying degrees they sensed this—no replacement postulate, even one of immense intuitive appeal, could possibly serve, for no matter how preferable it was psychologically, it would be equivalent to Postulate 5 and therefore indistinguishable from it where it counted —logically.

Notes

1. *Poseidonios* was head of the Stoic school on Rhodes and a teacher of Cicero. His works have not survived, but some of their titles (which have) indicate the breadth of his interests: *Treatise on Ethics*; *Treatise on Physics*; *History of Pompey's Campaigns in the East*; *On the Universe*; *Introduction to Diction*; *On Emotions*; *Against Zeno of Sidon* (on geometry).

2. *see for ourselves*. We will establish shortly that Poseidonos' Postulate and Euclid's Postulate 5 are but different formulations of the same basic assumption. Some scholars feel that Euclid was aware of the formulation later used by Poseidonos but chose the formulation he did precisely because of its finite character.

3. *Neutral geometry*. The term seems to have originated with Prenowitz and Jordan in *Basic Concepts of Geometry* (Blaisdell, 1965). Most books use the term "Absolute geometry" introduced in 1832 by János Bolyai, one of the founders of non-Euclidean geometry.

4. *Wallis' Postulate*. Wallis actually made the stronger assumption that "to *every* figure there exists a similar figure of arbitrary magnitude." (Bonola, *Non-Euclidean Geometry* (1906; Dover reprint, 1955), pp. 15–17.)

5. *letter*. From Bonola, *op. cit.*, pp. 65–66.

Richard J. Trudeau was educated at Boston College and has taught at Stonehill College since 1970.

He writes, "My only other published mathematical work is *Dots and Lines* (Kent State University Press, 1976), which Dover republished as *Introduction to Graph Theory*, and an excerpt from *The Non-Euclidean Revolution* ("How Big is a Point?") that appeared in the *Two-Year College Mathematics Journal* in 1983 and won a Pólya Prize.

"I have an M. Div. from Harvard (1994) and have been a Unitarian Universalist minister for ten years. Since the publication of *The Non-Euclidean Revolution* most of my creative efforts have gone into sermon preparation (often my sermons involve a lot of scientific or historical research)."

10

Arithmetic in the United States

Until a few years ago, it was possible for someone to graduate from an American high school without ever having to confront algebra. Now more and more states are including algebra problems on the tests that students must pass in order to get a high-school diploma. Students, seeing this, will take a course in algebra, thus tending to make algebra instruction universal.

This is not a bad thing because, as all teachers of mathematics know, algebra is good for people. It forces them to concentrate on thinking rationally and shows the power of rational thought to solve problems. The general public seems to agree with teachers of mathematics about the value of algebra because there has been no outcry about its expansion. People feel that, even if they did not like the subject, it was in fact good for them.

Before algebra comes arithmetic. It may come as a surprise that arithmetic instruction has not always been part of the education of American children, but that is the case. How, you may ask, could society possibly function if not everyone knew arithmetic? Well, it did, and not too badly. The feeling in the eighteenth century was that arithmetic, though necessary for trade, could safely be omitted from general education. Those who really needed it could learn it on the job. In some places there was opposition to including it in the course of instruction.

Also, as we will see in the following selection, instruction in arithmetic, when it was offered, was of the memorize-these-rules variety, which is not very good for people. Teachers of mathematics are well aware of this, which explains the perennial claim of mathematics educators that the last generation was taught wrongly, by rote, but this generation will teach all students to understand. The method that will do this remains undiscovered, probably because it does not exist. Try as we will to impart understanding, some people just will not understand. However, most of them can, if we insist, memorize rules and apply them. So, to achieve measurable results, the temptation is ever present to let understanding go by the boards and make sure that the students can do something. There is thus a constant slippage, which explains the periodic reforms and revolutions. Mathematics education is like the snail in the well, climbing up three feet every day but sliding back two feet every night, except that the education snail slips back three feet between reforms. We must have continual reform so as to stay in the same place.

It is instructive to see where we have been, which is why this selection is included here. If nothing else, history keeps us from assuming that the world has always been more or less what it is like now, and that it will continue to be so.

A Calculating People

Patricia Cline Cohen (1982)

Republican Arithmetic

In 1801 the editor of a fledgling Philadelphia journal of literary miscellany wrote a brief article in which he described, in a light and humorous style, his aversion to arithmetic. "Oliver Oldschool"—the pen name is significant—confessed to readers of *The Portfolio* that he had never studied much arithmetic because he considered it a shopkeeper's business: "It always appeared to me that a scholar could attain the object of his mission to the university, without any assistance from the first four rules [of arithmetic]." His college friends had ridiculed him, but in rebuttal he cited such diverse authorities as Shakespeare, Dr. Johnson, and the Calvinist John Knox to show in what low esteem past worthies had held arithmetic. He readily dismissed a current theory of the value of arithmetic: "We are magisterially told that this study, of all others, most closely fixes the attention. An argument shallow, untrue, and easily vanquished. Any object, that engrosses the mind, will induce a habit of attention."

The significance of Oldschool's objection to arithmetic lies in the fact that an antiarithmetic argument with a long and serious tradition had by 1801 come to be presented as an amusing filler piece. It was undoubtedly Oldschool's personal view that too close an attention to numbers was demeaning, that "the influence that arithmetical minutiae has gradually obtained over the heart" was deplorable; yet he titled his essay "Farrago No. V" (farrago means "hodgepodge"), and he led off with a bit of entertaining doggerel that detracted from the potential seriousness of his message:

> Our youth, proficients in a noble art,
> Divide a farthing to the hundredth part.
> Well done, my boy, the joyful father cries,
> Addition and Subtraction make us wise.

Oldschool was treading this thin line between serious cultural commentary and trivial amusement because at the beginning of the nineteenth century arithmetic itself was undergoing an important transformation. Since the seventeenth century, arithmetic had been thought of as a body of numerical rules and definitions that applied primarily to commercial life; arithmetic had no other uses, in the popular view, and only certain groups of people needed to know it. But at the end of the eighteenth century arithmetic began to lose its commercial taint because more applications were evident and because commerce itself was no longer restricted to a small segment of the American population. At the same time it was beginning to be argued that arithmetic had the power to improve the logical and rational faculties of the mind and was therefore particularly worthy of attention in a republican education. Oldschool was giving voice to rather old-fashioned arguments against the study of mathematics, and he must have recognized that his view was increasingly untenable. He ended his article abruptly, alluding to further criticisms he refrained from making because "the world, quoth Prudence, will not bear it," for "tis a penny-getting, pound-hoarding world."

Oldschool's antimodernism probably evoked agreement in some readers, a blush of guilt in others, but derisive scorn in the majority. In the first two decades of the nineteenth century, the American public's sensitivity to quantified material was considerably heightened by economic, political, and social changes that were making numbers an integral part of life. The commercial revolution stimulated reckoning skills as it pulled more people into a market economy. The political revolution that mandated the pursuit of happiness as an important end of government found its proof of the public's happiness in statistics of growth and progress. The proliferation of public schools, designed to ensure an educated electorate, provided a vehicle for transmitting numerical skills to many more people. These were the years when arithmetic became democratized—and also sex-typed. The history of the ways of teaching arithmetic, from the private tutors of colonial cities to the "mental arithmetic" theory of the 1820s, illuminates deep shifts in American attitudes toward commerce, democracy, and gender relations.

In seventeenth-century America there were few formal institutions of learning, and what few there were generally ignored arithmetic. The Puritans' remarkable provision for both primary and advanced education, extending from town schoolhouses to Harvard College, did not embrace the subject. But then, the men who designed Puritan schooling were themselves products of the English system, which considered arithmetic a vulgar study not properly part of a high-quality education. The ultimate end of a Puritan education was theological, not mercantile or scientific. Puritan children were to know how to read the Bible, and Puritan ministers endeavored to learn how to interpret and translate theological imponderables. There was no need to waste time learning much mathematics. Indeed, the 1648 inaugural rules for a grammar school in New Haven specifically directed that the pupils be taught to read, spell, write and "cypher for numeracion, & addicion, noe further." Perhaps the English mathematician John Arbuthnot was entirely correct when he charged that the typical American living in barbarous circumstances could not reckon over twenty. Most of them were farmers in a subsistence economy; they would, indeed, have had little use for arithmetic beyond the simple addition of small numbers.

In the eighteenth century, basic arithmetic training in the schools barely improved. In New England, which boasted the most advanced system of public education in all the colonies, local primary schools still did not teach arithmetic as part of their regular course of instruction. But the grounds for excluding it had now shifted. Whereas arithmetic was earlier ignored because it was not thought valuable, in the mid-eighteenth century it was regarded as too difficult for children younger than ten or twelve to study. However, since children older than twelve rarely continued to attend school, the result, a low level of numeracy, was still the same.

On occasion a group of older boys might prevail upon a schoolmaster to offer special arithmetic instruction. The custom was to hold the class in the evening, probably because, during the day, the youths were engaged in work and the schoolmaster was tied up teaching younger pupils to read and write. The popular English texts by Cocker and Hodder generally provided the structure for these evening classes, thus ensuring that in America, as in England, arithmetic would be identified with commerce. To what extent the youthful scholars were serious about their study of arithmetic was a question a schoolmaster had to resolve before he agreed to take on the extra work of evening lessons. One Connecticut teacher recalled that in the late eighteenth century he held evening arithmetic classes "by particular request of the pupils: but I always avoided them when I could ... because I had reason to doubt both the motives of those pupils who were the loudest petitioners on the subject, and their practical utility whenever they were permitted." It is a revealing commentary on the attractions of home and hearth in this period that boys would ask to have an evening school, where they would pursue a subject of dubious utility in their lives, from texts that certainly were not entertaining and were probably frustrating, and from a schoolmaster suspicious of their motives. Another district schoolteacher in early nineteenth-century New England recalled that that there had been "a most determined opposition" to the idea of transferring arithmetic to the day school; unfortunately, he did not elaborate on who opposed it and why. Was it the parents, who felt, perhaps, that arithmetic should remain an elective subject? Or was it the adolescent boys, who would lose their chance for an evening out?

Boys who went beyond the district school, into Latin grammar schools or, in the late eighteenth century, into the academies, encountered basic arithmetic in their required programs. But the traditional English gentlemanly disdain for the subject survived in muted form in America. Isaac Greenwood, first Hollis Professor of Mathematics at Harvard, published an elementary arithmetic text in 1729 for "Persons of some Education and Curiosity," yet he felt obliged to apologize to his elite readers: "It cannot be thought an unprofitable task for a Gentleman, especially of Curiosity and Learning, once in his Life to pass through the Rules in this Art." Apparently the teachers in the Latin schools agreed that it was not prohibitively vulgar and set their older pupils to work at elementary arithmetic. Benjamin Franklin's proposed curriculum for a Philadelphia academy in 1749 was more boldly utilitarian and included "Arithmetick, Accounts, and some of the first Principles of Geometry and Astronomy" near the top of his list of useful subjects." Daniel Webster was first introduced to arithmetic when he entered the Phillips Academy at Exeter in the 1790s, having learned only reading and writing in his district school in New Hampshire.

The inclusion of arithmetic in a Latin school or academy curriculum did not necessarily mean that the students actually learned much arithmetic. Caleb Bingham, an education reformer of the mid-nineteenth century, charged that an academy teacher of the 1790s, one of the most respected in Boston, had no idea how to do simple calculations. He took problems out of the copybook he had used as a boy and set them in the books of his beginning students, who were eleven years old: "Any boy could copy the work from the manuscript of any other further advanced than himself, and the writer never heard any explanation of any principle of arithmetic while he was at school. Indeed, the pupils believed that the master could not do the sums he set for them." Bingham added a predictable story about a boy who contradicted a sum in the master's book and incurred great displeasure, until the teacher reasserted his authority by altering the terms of the problem.

Copybooks were widely used in the eighteenth century as substitutes for textbooks, ever in short supply. The teacher dictated rules and problems to be copied into blank books, and the students worked out the answers with varying degrees of individual effort. A collection of sixty surviving arithmetic copybooks dating from the period 1739 to 1820 affords a glimpse of student interaction with the fundamentals of mathematics. Most of the books are neat and orderly, showing that penmanship as well as accuracy in problem-solving mattered a great deal. A few schoolboys succumbed to the temptation to doodle in the margins or to write names endlessly over the flyleaf, but most look like what they in fact became: books written to be saved a long time, for later reference. (Some contain other information, of the character of a permanent record: dates of children's births, commercial accounts, names of debtors and creditors; one writer even listed the dates on which various cows of his "went to bull" in the summer of 1775.) Those that contain dates reveal that the student completed his course of study in the subject in one or two years, and, if steadiness of hand is any guide to a child's age, it appears that all these scholars were older than ten when they began their study. (One recorded his age—thirteen.) The books trace a standard progression from addition, subtraction, multiplication, and division to the Rule of Three, single and double, inverse and direct. A few repeat these rules with fractions and decimals, and a very few go beyond, to simple measurement of angles. The most advanced copybook covers practical trigonometry and geometry as applied to problems in surveying and navigation.

Merely to list the subjects contained in the copybooks does not reveal the sufferings endured by these students. Types of problems that would be relatively easy for a high-school student of today take on nightmarish qualities when translated into the measurements and currencies of the eighteenth century and tackled with eighteenth-century methods. Arithmetic was a commercial subject through and through and was therefore burdened with the denominations of commerce. Addition was not merely simple addition with abstract numbers, it was the art of summing up compound numbers in many denominations— pounds, shillings, pence; gallons, quarts, and pints (differing in volume depending on the substance being measured); acres and rods, pounds and ounces (both troy and avoirdupois), firkins and barrels, and so on. Eighteenth-century copybooks show that students had to memorize all these tables of equivalences before embarking on the basic rules and operations. A large chunk of time was spent on a subject called "reduction": learning to reduce a compound number to its smallest unit to facilitate calculations. Students would practice on questions like How many seconds since the creation of the world? How many inches in 3 furlongs and 58 yards? This would prepare them for more advanced problems, such as What will ten pairs of shoes cost at 25 s. 6 d. a pair?

Eighteenth-century methods of problem-solving further confounded all but the best students because they deliberately relied on memory, not on understanding. The English assumption that arithmetic was too difficult to explain persisted in America; printed texts and students' copybooks contained a plethora of rules (sometimes in verse to aid memorization) to cover every conceivable sort of problem. The student's task was to match the problem to the correct rule and grind out the solution mechanically. Independent thought was not encouraged by this system. There were rules of simple addition, addition of compound numbers, and addition of fractions, often in widely separated chapters of a text, with no clue to the fact that they are all variations of the same operation. There were the Rule of Three Direct, the Rule of Three Inverse, the Rule of Fellowship, the Rule of Interest, and dozens more.

Here is a problem in "discount": "What is the present worth of $5,150 due in 44 months at 8% per annum and 1% for prompt payment?" The modern student would be likely to set up an equation that expressed the relation of all numbers to the unknown, the present value. The hapless copybook writer had no algebra or experience with equations to help him think logically about the problem. Instead, a rote rule in discount was applied and the answer ground out, with no sense that this type of problem might in any way be related to lots of other problems not involving interest and present values.

The famous Rule of Three illustrates nicely the potential for confusion in eighteenth-century arithmetic. After the rules of addition, subtraction, multiplication, and division of whole numbers, the Rule of Three was the next step and the most basic operation of business—so basic that it was often called the Golden Rule. Many students' entire arithmetic education ended with the Rule of Three. "Given three parts, to find the fourth," is the way the rule was often stated. If 7 yards of cloth cost 21 shillings, how much do 19 yards cost? The Rule of Three instructed the student to set down 7:21::19 on paper and then multiply the middle times the last number and divide the product by the first. Writing the numbers in the wrong order was a major source of grief to students. But the real problem with this rule was that the first step of the solution leads to a meaningless number: 21 shillings by 19 yards equals 399 units of what? It is only with the second step, dividing 399 by 7 yards, that the calculation makes any sense. A more logical way to do the problem is to figure out what one yard costs and go on from there.

This pitfall in the Rule of Three made it hard for students to realize when they had gone awry in a problem. Here is a typical word problem, typical in its complexity and in its use of current events to suggest the utility of arithmetic:

> Suppose General Washington had 800 men and was supplied with provision for but two months, how many of his men must leave him, that his provision may serve the remaining five months?

In this particular case the student mechanically applied the Rule of Three, writing 2:800::5 and then dividing 5 into 2 × 800 to get a final answer of 320. Now, 320 is the number of men who can be fed for five months, not the number who must leave; so Washington's troops would have gone hungry had this schoolboy or his master been in charge of provisioning. One suspects that a real military man would have put the question in a different way: How much food must I commandeer to feed my 800 troops for three more months? But then, problems were not always totally practical: "How many minutes [were there] from the commencement of the war between America and England, April 19, 1775, to the settlement of a general peace, which took place January 20, 1783?"

If these copybooks accurately reflect the substance of arithmetic instruction in academies and grammar schools, then it is easy to see why colonial colleges in general regarded arithmetic as a vulgar subject that could safely be ignored by men heading for a clerical or legal calling. But, gradually, colonial colleges in the eighteenth century became more rigorous in their mathematics curricula and requirements for admission. Although Harvard had established a mathematics professorship in 1726, it still relegated arithmetic, geometry, and astronomy to a mere two hours a week in the senior year. Sometime before the 1780s arithmetic became a first-year subject, leaving room for algebra, geometry, and higher branches in the following years. Yale offered arithmetic as a senior course as early as 1716, and in the 1740s, under the presidency of the mathematically-minded Thomas Clap, it made arithmetic a first-year subject and added algebra and geometry to the curriculum. A few advanced students even studied conic sections and fluxions—calculus—in their senior year. In 1745 Yale daringly began to require a mastery of basic arithmetic before admission to the college, as did Princeton in 1760; but Harvard waited until 1802 to institute a similar rule. The mathematics programs at William and Mary, the University of Pennsylvania, and Dartmouth were strong by the end of the eighteenth century, but, like Harvard, they all began with a quick course in the first four rules. The general assumption was that entering students had a fairly low or even nonexistent acquaintance with basic arithmetic.

Although public schools and academies were less than thorough in their arithmetic instruction, there was a second and much more common way to gain mathematical knowledge in the cities of colonial America: from mathematical practitioners and other private masters in day and evening schools. Throughout the eighteenth century, newspapers carried advertisements for special courses in writing and arithmetic, in the tradition of the English petty school. Sometimes the notices promised great scope and advanced mathematical learning, as in this item from the *Boston Gazette*:

> At the House formerly Sir Charles Hobby's are taught, Grammar, Writing after a free and easy manner, in all the hands usually practiced, Arithmetick Vulgar and Decimal in a concise and practical Method, Merchants Accompts, Geometry, Algebra, Mensuration, Geography, Trigonometry, Astronomy, Navigation, and other parts of the Mathematicks, with the use of the Globes and other Mathematical Instruments, by Samuel Grainger.

Another advanced teacher was Isaac Greenwood, who set himself up as a private master after he was removed from his position at Harvard for intemperance in 1738, as did Nathan Prince, another mathematician of repute, who lost the competition for Greenwood's chair to John Winthrop. But more often the advertisements for lessons offered less impressive programs and were placed by men who had no connection to the scientific elite in the colonies. In South Carolina Stephen Hartley offered reading, writing in several hands, and "Arithmetick, in all its Parts, Merchant's Accompts, or the Italian Method of Bookkeeping" to Charleston residents in June of 1744; three months later George Brownell and John Pratt placed a notice offering the same subjects plus dancing; and the competition was joined by a husband-and-wife team, who included needlework and drawing for young ladies in their program.

It is evident that these evening schools filled a need for basic education that was being met in no other way. The teachers were ordinarily men who had probably learned the material from a private master themselves. Teaching was something they could always fall back on when other types of employment did not work out, as is evident in this plea by a seemingly multitalented Thomas Smith of Virginia:

> To the PUBLIC. A CLERGYMAN of character, of the church of England, addresses himself to the churches of all his Majesty's plantations, that if they want a sober young man for their Minister, on trial, on reasonable terms: Or, he proposes to teach young Gentlemen and Ladies French, Latin, Greek, and English, book keeping by double entry, algebra, ge-

ometry, measuring, surveying, mechanics, fortification; gunnery, navigation, and the use of the globes and maps.

So, although arithmetic was not taught in the district schools and was given
only cursory attention in higher schools, there were other avenues for acquiring
mathematical skills. Indeed, it would be very surprising if this society, whose
elite class derived its status from commercial activity, had not created some
institutional or formal method for maintaining the requisite numerical skills.

The advertisements for evening schools and writing schools are only the
most visible evidence of the ways by which arithmetic knowledge was conveyed. Less formal (and therefore less visible to historians) was the education
conveyed from master to apprentice and from parent to child. Apprenticeship
laws in the seventeenth century specified that masters be directed to teach their
charges to read and write; in the eighteenth century in several colonies the law
was enlarged to include "ciphering as far as the rule of three." Probably not
a few masters in urban areas sent their apprentices to the evening schools to
fulfill this part of their obligation. As for the instruction that passed from parent
to child, little trace remains. Common sense suggests that whenever a parent
knew some arithmetic and also found it useful in his vocation, he would teach
it to his child if no institution did. Or, lacking time but not money, a parent
could hire a tutor: "Wanted soon: A TUTOR for a private family, who among
other things, thoroughly understands the mathematics." Finally, we should not
overlook the possibility that unusually able men could learn arithmetic on their
own. Several of the most famous colonial scientists were entirely self-taught:
David Rittenhouse, Nathaniel Bowditch, and Benjamin Banneker. Benjamin
Franklin had failed to prosper under a teacher, but at the age of sixteen he
"took Cocker's book of arithmetic, and went through the whole by myself with
the greatest ease"—a truly commendable feat, considering the obfuscating nature of Cocker's text. Clearly, it was possible in eighteenth-century America to
learn the rudiments of arithmetic if one needed it in life.

That there were, then, a number of different avenues for learning arithmetic should not obscure the overriding fact that arithmetic instruction remained
inflexible and barren of innovation throughout the eighteenth century. It was
recognized to be a troublesome subject, but no one made an effort to simplify
it. The response instead was to postpone teaching it until a child was ten or
twelve (or older). Only those with a future in commerce made a thorough study
of it, and, even then, there were shortcuts to reduce reliance on calculations and
ways to make up for poor training.

For those who found that they lacked quickness and skill, there were ready
reckoners to provide a handy method of computation. Just like English reck-

oners, these were books consisting of hundreds of tables listing multiples of various unit prices, so that one could look up the price of any amount of a commodity.

> Although this book be chiefly calculated for the use of people unacquainted with figures, yet there is no doubt of its being of advantage, to those who may be perfect in arithmetic; as mistakes will frequently escape the readiest Penmen, when in the hury [*sic*] of business.

In addition to the unit-price tables, ready reckoners provided other sorts of tables that reflected the changing needs of the public. A 1774 edition compared the value of Pennsylvania and New York currency in terms of gold and silver, an appropriate nod to the increasing interdependency of the colonies; a 1789 edition presented "A Scale of Depreciation, for the Settlement of old debts" for Pennsylvania and Maryland; and a 1794 edition provided a table converting cents to pounds, shillings, and pence, "very useful to all Persons who may be concerned in the Duties established by the Revenue Laws of Congress."

An early and ambitious American version of the reckoner was a reference work written in 1731 by Thomas Prince, a Boston clergyman; he called it *Vade Mecum for America: or a Companion for Traders and Travellers*. Prince grandly claimed that his book was the first of its kind and of such "great advantage, that we need not speak a word in its Commendation." He combined the price tables of the ready reckoner with the sort of miscellany more often included in almanacs. After more than 150 pages of tables showing the value of any quantity of a commodity, there followed a table of interest at 6 percent for a variety of principals; a list of counties and towns in New England and the court dates of colonies as far south as Virginia; the dates of Quaker meetings and fairs; an account of the principal stages along the roads from Boston and the distance between each tavern, down to the nearest half-mile; and a table of weights and measures in all their eighteenth-century complexity. This much was undoubtedly of use to readers, but, from there on, Prince got carried away with his compilation of quantitative information. He provided a chronology of the kings and queens of England, noting dates of birth and death, the date of the beginning of each reign, and then the length of each reign, figured in years, months, and days. The curious reader next encountered an alphabetical list of all the streets in Boston, followed by a neat summary: 60 streets, 41 lanes, and 12 alleys. It is hard to imagine what practical use Prince thought this information might have. Here we encounter an extreme of the quantitative mentality in the eighteenth century: a pure love of counting and computation. Prince, in his position as Congregationalist minister, followed with interest the patterns of vital events among his parishioners and on occasion brought demography and political arithmetic into his sermons. Evidently the same spirit of data collection carried over into his *Vade Mecum*. American ready reckoners after Prince's were limited to tables of prices and interest; the courts and roads, fairs and meetings, were left to the almanac trade, the weights and measures to arithmetic books. No one else bothered to disseminate information about the duration of monarchs or the number of streets, lanes, and alleys in a town—eloquent testimony to the low level of interest such quantitative trivia commanded in the eighteenth century.

The appeal of basic ready reckoners persisted in the eighteenth century because the tables facilitated commercial transactions in a society where few were adept at juggling pounds and shillings along with ells and firkins. But by the second decade of the nineteenth century the use of reckoners had died out, for two reasons. The gradual shift to a decimal monetary system, begun in 1792, greatly simplified price calculations; proponents of the new federal money claimed that they were democratizing commerce by putting computation within the reach of nearly all. At the same time, the self-consciously utilitarian spirit of the new nation invaded education and elevated arithmetic to the status of a basic skill along with reading and writing. Decimal money and arithmetic education were justified as fruits of republican ideology; numeracy was hailed as a cornerstone of free markets and a free society.

Thomas Jefferson proposed that the United States adopt decimal money in 1784, when the Continental Congress was struggling to create order amidst the financial chaos of the Confederation period. As obvious as the virtue of a decimal money system may seem today, it was not so obvious in the 1780s. In fact, Jefferson's plan to decimalize the currency was not the first standardization plan that came before the Continental Congress. In 1782 Gouverneur Morris, assistant to financier Robert Morris, drew up a proposal to base the currency on the unit of silver that was the lowest common denominator of all the various states' pennies. This he determined to be one quarter-grain of silver, and from that base a standardized dollar would contain 1,440 units, a crown, 1,600 units. The virtue of this system was its compatibility with the state coins; its drawback was that it created terrifying prices. As Jefferson pointed out, a loaf of bread costing one-twentieth of a dollar would instead cost 72 units; a pound of butter, 288 units; a horse, 115,200 units; and the public debt—in 1782 a matter of mounting concern—would be too shocking to contemplate.

Congress tabled Morris's report for two years, then referred it to a committee, of which Jefferson was a member. The committee then proposed a new system of dollars, divided into dimes and cents, and established a fixed ratio of gold and silver values to standardize it. The new plan was accepted by Congress, and in 1792 the Mint, established at Philadelphia, began to make pennies and half-pennies. By 1796, with the issue of dimes and halfdimes, Jefferson's prediction about the superior ease of reckoning in decimals could be tested:

> The facility which this would introduce into the vulgar arithmetic would, unquestionably, be soon and sensibly felt by the whole mass of people, who would thereby be enabled to compute for themselves whatever they should have occasion to buy, to sell, or measure, which the present complicated and difficult ratios place beyond their computation for the most part.

Before the facility could be sensibly felt, however, people would have to be instructed in the use of the new money, and several books and pamphlets appeared in 1795 and 1796 to that end. Some were mere broadsides—tables of exchange rates between old and new currency, suitable for nailing to the wall in stores. Others were complete arithmetic texts that recast the first four rules with examples in federal money, specifically for the benefit of all who wanted to join the "rapid strides towards commercial grandeur" then under way but who had felt excluded and confused by the intricacies of the old system of money. *The Assistant, or A Treatise on the Currency of the United States* (1796), addressed itself to "Traders, Mechanics and young persons" and gave rules and examples "sufficient to teach the smallest capacity, without the assistance of an instructor." In *An Introduction to Arithmetic for the Use of Common Schools* (1796), author Erastus Root specified his audience as future farmers and mechanics and made explicit the patriotic interconnections between common arithmetic, decimal money, and republican government:

> ... it is expected that before many years, nay, many months, shall elapse, this mode of reckoning will become general throughout the United States.... Then let us, I beg of you, Fellow-Citizens, no longer meanly follow the British intricate mode of reckoning.—Let them have their own way—and us, ours.—Their mode is suited to the genius of their government, for it seems to be the policy of tyrants, to keep their accounts in as intricate, and perplexing a method as possible; that the smaller number of their subjects may be able to estimate their enormous impositions and exactions. But Republican money ought to be simple, and adapted to the meanest capacity.

Decimals had been studied by mathematicians for two hundred years, but America was the first country to put them to practical use, Root boasted, drawing a parallel to the ancient Tree of Liberty, existing eternally but blossoming only in the hospitable American environment. This line of thinking unavoidably required a comment on France, which had followed America in both liberty and decimals. By 1796 France had adopted decimal coins, metric weights and measures, and a revamped calendar of ten-day weeks, which afforded Root an opportunity to make a pun: just as they had gone too far in republicanism, "so have they stretched decimal simplicity beyond its proper limits, even into decadary in fidelity."

The introduction of decimal money removed a significant obstacle to the study of arithmetic, inasmuch as arithmetic was still taught and thought of primarily in a commercial context. Dollars, dimes, and cents were certainly easier to figure with than pounds, shillings, and pence; they alleviated a lot of the pain associated with denominate arithmetic. The new system of money also required the publication of new arithmetic texts, ones appropriate to American life. Cultural chauvinism, already set in motion by the Revolution, led to a decline in the pirating of English texts and to their replacement by American texts written by Americans. With the introduction of federal money, the American authors had reason to begin to alter the substance of arithmetic texts; no longer did strict adherence to tradition prevail.

A second development in the late eighteenth century further eroded the traditional memory-based version of arithmetic. The arrival of republican government provoked intense speculation about the form and content of education appropriate to a republic. The consensus was that education should be widespread and that it should teach future citizens to think. Several leading figures in the 1780s and 1790s wrote essays describing the sort of education best suited to the rising generations in the new United States, and there was universal agreement that arithmetic was essential and should be elevated to the level of a basic skill. In its new role as a required subject, arithmetic would promote the spread of the commercial frame of mind and at the same time foster a citizenry able to reason clearly. In *Notes on Virginia* (1782) Thomas Jefferson proposed a system of local schools, open to all, that taught reading, writing, and "common arithmetic"; the best of the pupils would have the opportunity to study "the higher branches of numerical arithmetic" at the next step up, the grammar school. Benjamin Rush, the Philadelphia physician, advocated that free district schools teach reading, writing, and the "use of figures." Noah Webster agreed that the yeomanry in district schools should learn about commerce and money, but he kept to the traditional view that boys at age ten were still too immature to reason about abstract mathematics. Two other men, a journalist and an educator, who shared a prize from the American Philosophical Society in 1797 for their essays on education, both matter-of-factly included arithmetic in their proposed primary-school curricula.

While a new consensus was emerging on the value of arithmetic, none of these early advocates of republican education considered whether the method of teaching it ought to be altered. In the 1790s the subject was frozen in the structured texts of Cocker, Hodder, Dilworth, and their imitators, though their complexity had been ameliorated somewhat by the substitution of decimal money in the examples. Arithmetic was still a laborious study, dependent on rules, catechisms, and memory work. Thus it had been since the seventeenth century, and no one had stopped to wonder whether there might be a better pedagogic approach. In the process of making arithmetic a required subject for all young pupils, educators finally had to conclude that the traditional method of instruction simply did not work very well. A self-selected group of fifteen-year-old boys could wrestle with Cocker and finally make sense of it, but the average eight- or ten-year-old needed explanation and help that schoolmasters of the young republic were not well trained to give.

The problem is illustrated by what happened in Boston, which in 1789 passed its own Education Act, a part of which specified that boys aged eleven to fourteen were to learn a standardized course of arithmetic through fractions. Prior to this act, arithmetic had not been required in the Boston schools at all. Within a few years a group of Boston businessmen protested to the School Committee that the pupils taught by the method of arithmetic instruction then in use were totally unprepared for business. Unfortunately, the educators in this case insisted that they were doing an adequate job and refused to make changes in their program.

In general it appears that, although the value of arithmetic knowledge was finally recognized, most communities lacked the skilled teachers and comprehensible texts needed for transmitting it. In the district schools in New England arithmetic was repeatedly judged to be in "a very low state." Benjamin Latrobe, a prominent architect and engineer, pinpointed an obstacle that had yet to be overcome:

> Arithmetic is generally a heavy study to boys, because it is rendered entirely a business of memory, no reasons being assigned for the rules. A schoolbook of arithmetic accompanied with demonstrations is much wanted. We do boys from seven to fifteen years old a great injustice in supposing they cannot reason.

Here Latrobe touched on a claim that had long been accepted for more advanced mathematics: that the ability to reason carefully and logically was associated with mathematical skill. However, elementary arithmetic had been so closely allied to commercial exchange since the seventeenth century that it had been overlooked as a purely intellectual exercise for the mind. With the changing attitudes toward public education and the desire for a populace well attuned to reason and logic as a solid foundation for republican government, there emerged the idea that even very basic arithmetic could help train men to think well. In the summer of 1788 George Washington expressed this idea in a letter to Nicolas Pike, the author of the first purely American elementary arithmetic text to be published in the new republic: "The science of figures, to a certain degree, is not only indispensably requisite in every walk of civilised life, but the investigation of mathematical truths accustoms the mind to method and correctness in reasoning, and is an employm. peculiarly worthy of rational beings." Next, Washington rather cloudily affirmed the psychological satisfaction inherent in arithmetical precision: "In a cloudy state of existence, where so many things appear precarious to the bewildered research, it is here that the rational faculties find a firm foundation to rest upon."

Washington's youthful experiences as a surveyor supported his belief that mathematics could be the cornerstone of rationality. Jefferson, who had also been a surveyor, praised the numerical art as delightful in itself and as a good preparation for other intellectual endeavors: "The faculties of the mind, like the members of the body, are strengthened and improved by exercise. Mathematical reasoning and deductions are, therefore, a fine preparation for investigating the abstruse speculations of the law." Many journal articles of the early nineteenth century repeated the notion that arithmetic had a double function: it was useful for practical life and it was an exercise for the mind. Not everyone agreed with this view, but opponents did not flock to the literary and education journals to express antimathematical sentiments. An 1821 article in the North American Review elaborated on objections to the study of mathematics in order to refute them, and perhaps this paragraph can be taken as a fair representation of the antimathematics view:

> It is sometimes objected to the study of mathematics, that it contracts the mind, and, by circumscribing its view, opposes the exercise of invention; that it tends to form a mechanical and skeptical character, rendering the mind incapable of comprehending an extensive subject, and insensible to those nice shades of evidence, and unsusceptible of that accurate perception of beauty and truth, so requisite to quick and fair judgment in matters of taste and morals. This charge, if well founded, would be sufficient to prove this study to be dangerous; and we have no doubt that a belief

more or less confident, of its justice, still operates on many persons in prejudice of mathematical pursuits.

Perhaps so, but the "many persons" who adhered to this prejudice did not find the courage to voice their objections to the growing acceptance of mathematics education. Oliver Oldschool's "Farrago No. V" stands out as an unusual editorial comment swimming against the tide.

With all the new emphasis on arithmetic as a foundation for rational thinking, it became especially clear that eighteenth-century texts, based on memory work at the expense of logic, had to be completely revised. A revamped arithmetic that fostered rationality through mental discipline had to be simplified so that young children could not only learn it but understand it. Beginning in 1800 several authors tried by piecemeal improvements to overcome the inadequacies of the traditional texts. David Cook, Jr., of New Haven, decided that doing away with fractions altogether would lighten the burden for students and make arithmetic more attractive. He proposed substituting decimals:

> By this method of simplifying fractions, the mind both of the learner and practioner [*sic*] are freed from the heavy embarrassment of copying therein, such numbers of irregular tables, as they by every other method, are both obliged to learn and retain; by reason of which, multitudes of our fellow citizens, as well as the inhabitants of other countries, have been discouraged in the pursuit of Arithmetic, and thereby rendered incapable of performing those services either to their families or respective countries.

Seven years later, in 1807, a Philadelpia text by Titus Bennett tried to adapt arithmetic to the capacities of young children by simplifying the rules and shortening the examples; the use of federal money in all examples was the key element in his simplification. In 1811 a Massachusetts author predicted that what he called the three basic parts of arithmetic—numeration, operations with simple numbers, and operations with compound numbers—would soon be reduced to two when Americans abandoned the irregular system of weights and measures that formed the substance of compound arithmetic. What need have children to learn wine and beer measure? it was asked.

Postponing vulgar fractions and eliminating the bulk of denominate numbers streamlined the earliest lessons in arithmetic, making it feasible for five- and six-year-olds to concentrate on simple operations with simple numbers. There was still memory work, to be sure, but the memory was now employed in learning number facts, not tables of equivalences. Drill on the four basic operations became a major object of early instruction. Colonial copybooks and texts give no sense that students were ever drilled on basic computations. Between 1800 and 1830, games and prizes for correct recitations of the multiplication table became common. One author recommended that the ultimate goal of drill should be to bypass words and achieve instant recognition of the sum or product of two numbers. That is, instead of learning a singsong "nine times seven is sixty-three" students should learn to glance at a 9 and a 7 and instantly think 63; translating into words was an extra and unnecessary step

Close on the heels of simplified arithmetic came the idea of teaching beginning arithmetic with tangible objects like counters and bead-frames. The earliest proponent of this was Samuel Goodrich, author of *The Child's Arithmetic*, who in 1818 argued that learning by rules and rote actually prevented children from comprehending arithmetic. Children should instead discover rules by manipulating tangible objects. The idea here was to get children to understand in a physical way what goes on in addition or multiplication (really successive addition) long before they moved on to abstract numbers. Goodrich's idea was taken up and elaborated within a few years by a young Harvard-trained mathematics major who called the new system "mental arithmetic."

"Mental arithmetic" swept the field of mathematics education in the early 1820s. The conventional notion that arithmetic was a memory-based subject fit only for mature minds and chiefly a preparation for business was completely overturned. The young Harvard graduate who inaugurated the revision was Warren Colburn, who in 1821 published *First Lessons, Or Intellectual Arithmetic on the Plan of Pestalozzi*. It contained no rules, no memory work, and was intended for four- and five-year-olds. The *North American Review* prophesied, "We have no doubt that Mr. Colburn's book will do much to effect an important change in the common mode of teaching arithmetic." By 1826 Colburn had published a second edition, a sequel (*Arithmetic upon the Inductive Method of Instruction*), and an algebra text constructed on the same principles. Beginning in 1826, with the first issue of the *American Journal of Education*, reports came flooding in from schools in New England and New York about the great success of Colburn's method.

There were really two related but distinct concepts working behind Colburn's "mental arithmetic." The first was implied in the name: arithmetic done in the mind, without pencil and paper. Colburn proposed to teach very young children to grasp the idea of number by first using tangible objects and then proceeding carefully—and without the use of written symbols—to abstract numbers. Early parts of his texts spelled out numbers as words, as though the arabic symbols were an obstacle to thought: "fifteen and three quarters" made more sense to a beginning student than "15-3/4," according to Colburn, a notion that underscores the distance between literacy and numeracy in the pre-mental-arithmetic era. Colburn believed that children could develop their own calculative techniques, unhampered by conventional ways of working problems, without the benefit of rules and, apparently, even without the benefit of arabic numerals. Once students developed facility with mental arithmetic, they could then move on to written arithmetic.

The second concept implicit in Colburn's system was inductive reasoning. At the heart of his method was the notion that students would discover the basic rules of arithmetic for themselves by working carefully chosen problems. Set a student to an addition problem, he advised,

> without telling him what to do. He will discover what is to be done, and invent a way to do it. Let him perform several in his own way, and then suggest some method a little different from his, and nearer the common method. If he readily comprehends it, he will be pleased with it, and adopt it. If he does not, his mind is not yet prepared for it, and should be allowed to continue his own way longer, and then it should be suggested again.

The goal was to abandon slavish reliance on rules and memory work. Colburn declared (prematurely, as it developed) that the Rule of Three was dead, since it was worse than useless: it was an obstacle to thought. Children were finally encouraged to think through problems in proportion by using common sense.

The most grandiose claim for "intellectual arithmetic on the inductive plan" was that it permitted students to recapitulate the development of arithmetic as a system of thought; the excitement of discovery would fix the subject in their minds for the rest of their lives. Every child had the potential to be an original mathematical thinker, or so Colburn and his followers thought. In practice, it appears that fewer children than Colburn anticipated had the capacity to be that original; the journal articles of the period never mention the discovery of geniuses as a by-product of the inductive method. Instead, they emphasize the superior logic of Colburn's system as its chief merit. The *American Journal of Education*, for example, characterized the old method of arithmetic as synthetic, proceeding from rule to example, guaranteed to paralyze interest in mathematics, whereas the new method was said to be analytic, working from examples to rules. The teacher still, ultimately, handed down the rules, but now the students would understand them rather than memorize them. The author of "Errors in Common Education" called the new process a movement from the particular to the general, and he added, "I question, if there ever was a boy, who learnt to perform the simple rules of arithmetic, from the directions given in his book, unaided by some visible illustration of the process." But that was the mark of genius in Colburn's system—that within a few years it could seem to be the only way to learn arithmetic.

> The author has heard it objected to his arithmetics by some, that they are too easy. Perhaps the same objection will be made to this treatise on algebra. But in both cases, if they are too easy, it is the fault of the subject, and not of the book.

Why was mental arithmetic so immediately popular in the 1820s? The esteem for rapid mental calculation possibly reflects changed market conditions. If it is true that the market was more free and unfettered in the 1820s and 1830s than it had been before 1800, and if prices fluctuated more in response to short-run conditions, if there were more steps along the way adding value to a product and each step was itself a market calculation of costs and profits, then perhaps mental facility with figures was an adaptive skill. The rapid mental calculator had an advantage in the market, and adults, realizing this, began training their children to compete. But it might equally be true that mental arithmetic was invented by educators as a teaching device, to induce competition in the classroom (like the spelling bee, which we do not ordinarily envision as direct training for some specific adult activity).

The popularity of using arithmetic to foster inductive reasoning makes sense in view of the enormous prestige of inductive reasoning in the early nineteenth century. Here was the link between arithmetic and rational thinking. Students learned to move rigorously from facts to conclusions. This new version of arithmetic disciplined the mind and developed habits of precision, attention to detail and love of exact knowledge. All this was accomplished without abandoning the original mission of preparing youth for participation in the market, since the problems and examples that sharpened inductive thought were still largely drawn from the realm of commerce. Little children may have started out in Colburn, adding apples and oranges or mental quantities, but within a few years they were busy at work on interest questions, which occupied no little space in texts of the antebellum decades.

Colburn did not claim total originality for his method; he gave much credit to common sense.

> The manner of performing examples [i.e., problems] will appear new to many but it will be found much more agreeable to the practice of men of business, and men of science generally, than those commonly found in

books. This is the method of those that understand the subject. The others were invented as a substitute for understanding.

However, except for the title of his first book, Colburn did not give credit where it was really due: to Johann Pestalozzi, the Swiss educator who was revolutionizing educational theory with his view that children could take an active part in the learning process. And something more than common sense led to the widespread acceptance of Colburn's adaptation of the Pestalozzian inductive method. In the 1820s people became especially sensitive to the unique nature of childhood. "Mental arithmetic" appeared in the same decade as the introduction of infant schools, Sunday and common schools, a special juvenile literature, and pediatrics as a medical specialty. The same period saw the demise of the theory of infant damnation and the rise of a highly sentimentalized mother-child nexus. Children were no longer perceived as imperfect adults; they had a special psychology, and they passed through stages of intellectual development that were not simply deficient approximations of the adult stage.

Patricia Cline Cohen is chair of the history department at the University of California, Santa Barbara. Besides *A Calculating People*, she is the author of *The Murder of Helen Jewett: The Life and Death of a Prostitute in Nineteenth-Century New York* (1998).

11

The Moore Method

Robert E. Lee Moore (1882–1974) taught at the University of Texas from 1920 until he was forced finally to retire at the age of 86. As everyone knows, the only way to learn mathematics is to do it. Moore took this theorem and deduced a corollary, that students should be told nothing except the statements of theorems (and some definitions) and should then find all the proofs by themselves. The Moore method has not caught on generally, because it is hard on students, who prefer to be told things, and it is hard on instructors, who love to talk. However, it has its more or less fanatical adherents.

There follows a description of the method, and its creator, by Paul Halmos, a gifted teacher, writer, and lecturer. The book from which it is taken, *I Want to be a Mathematician*, contains reams of quotable material. Beyond the material on Moore I have included only one further excerpt, on what it takes to be a professional mathematician.

I Want to be a Mathematician

Paul Halmos (1985)

The Moore method

I can't remember when I became a Moore convert. I heard about R. L. Moore while I was still in graduate school; Felix Welch knew him at Texas and told stories about him, many legends were in general circulation, and I kept looking at Moore's blue Colloquium book hoping to make sense out of it. I never succeeded. I have always been fascinated by set-theoretic topology, and that's what the book seemed to be about, but the language was ugly and the theorems seemed much too special. Here is a sample (from p. 33 of the first edition).

> If M is a closed point set and G is an upper semicontinuous collection of mutually exclusive closed point sets such that every point set of the collection G contains a point of M and every point of M belongs to some point set of the collection G and K is a subset of M and W is the collection of all point sets of G that contain one or more points of K and no continuum of W contains a point of $M - K$ then W is a region with respect to G if and only if K is a domain with respect to M.

Moore felt the excitement of mathematical discovery and he understood the relation between that and the precision of mathematical expression. He could communicate his feeling and his understanding to his students, but he seemed not to know or care about the beauty, the architecture, and the elegance of mathematics and of mathematical writing. Most of his students inherited his failings as well as his virtues (diluted, of course); only the greatest, such as Wilder and Bing, could overcome the handicap of being a Moore student and become genuine mathematicians.

Moore insisted on using his own terminology, which was not that of the rest of the topological community, and he was inflexible about some of his mathematical attitudes. He wouldn't say "consider two points p and q…" for fear that in a degenerate case p could coincide with q, and then the number of points wouldn't really be two. He would never agree that the intersection of two sets is always a set, because, for him, there was no such thing as the empty set. These are forgivable quirks of language. Worse than that, however, he was usually intolerant about every part of mathematics other than his own: algebra and analysis were different subjects, competitors, enemies. That's bad: that made him a lesser mathematician than he had the talent to be, and it made his less talented students much less well educated and useful than they could have been.

He was a Texan, almost the fictional prototype. He spoke Texan, he was politically rigid, he had strong prejudices, he stood up when a lady entered the room, and (the story goes) he wouldn't accept students who were black, or female, or foreign, or Jewish. (At least a part of that is false: he had women PhD students, notably Mary Ellen Estill Rudin. So far as I know he did not have a black student.)

When I first heard about him, I was interested in the legend but emotionally neutral toward it. I have always had an ambivalent love-hate feeling to the

South (Texas included). Insofar as it represented old-fashioned gentility and courtesy, and soft speech and warm weather and mint juleps at the race track, I admired it and wanted to be a part of it—but when it represented prejudice and bigotry and hatred and violence I was repelled. The gentle southern belle, with her melodious accent and youthful grace I found attractive, of course; but the superannuated ex-belle, the middle-aged or elderly harridan, selfish, demanding, grasping, and domineering—from her I turned and ran as fast as I could. Moore, the educated well-spoken Texan mathematician extraordinary, was a hero of mine; Moore, the mathematically outmoded and ethnically prejudiced reactionary power, was a villain.

An effective hero, a productive villain, everyone must admit. He turned out a record-breaking number of PhD's in mathematics; they loved him and imitated him as far as they could. He did it by what has come to be called the Moore method. It is also called the Texas or Socratic or discovery or do-it-yourself method. I wish I could have seen it in action, instead of just hearing and reading about it, and, later, experimenting with modified versions of it. I got to know Moore slightly, and I did see the method in action, once, for one hour. When I passed through Austin one time, I screwed up my courage and knocked at Moore's office door. He received me politely and was willing to talk about his method. I was fond of the old man—he might have been bad and wrong in some ways, but he was human and strong, he was pleasant and interesting to talk with, and altogether I liked and admired him. (I enjoy remembering that when I took his photograph he insisted on combing his hair first—there was always a comb in his pocket—and he warned me that he would not smile—he explained that trying to smile made him look silly.) He seemed willing to put up with me. He invited me to stop by again, and I did.

I saw him that way, in Austin, perhaps four or five times. On one of those occasions he invited me to attend the first meeting of a calculus class he was teaching, and that was the only time I saw Moore in action. (I saw the movie about him, Challenge in the Classroom—in fact I saw it four times—but that's seeing him in action at a distance, not the real thing. Still, it's very good—I recommend it warmly.) He was good all right. He could draw the students out, he could keep things moving, he didn't bore anyone, his personality dominated the room.

What then is the Moore method? It begins with a preliminary interview: each prospective student must first go to see Moore and answer questions about his mathematical background. If he has already taken a course similar to the one he wants to be admitted to, if he knows too much, then the answer is an automatic no—he's out. A basic rule of the game is that all the players must start at the same place, and, in particular, must be untainted by the terminology, the notation, the methods, the results, and the ideas of others. A second rule is that the student, once he is admitted, must actively think, not passively read. Moore, so the story goes, would put students on their honor not to read about the subject while they were studying it with him, and, to lessen the temptation in their way, he would remove from the library all the pertinent books and journals. Now the class is ready to begin.

At the first meeting of the class Moore would define the basic terms and either challenge the class to discover the relations among them, or, depending on the subject, the level, and the students, explicitly state a theorem, or two, or three. Class dismissed. Next meeting: "Mr. Smith, please prove Theorem 1. Oh, you can't? Very well, Mr. Jones, you? No? Mr. Robinson? No? Well, let's skip Theorem 1 and come back to it later. How about Theorem 2, Mr. Smith?" Someone almost always could do something. If not, class dismissed. It didn't take the class long to discover that Moore really meant it, and presently the students would be proving theorems and watching the proofs of others with the eyes of eagles. One of the rules was that you mustn't let anything wrong get past you—if the one who is presenting a proof makes a mistake, it's your duty (and pleasant privilege?) to call attention to it, to supply a correction if you can, or, at the very least, to demand one.

The procedure quickly led to an ordering of the students by quality. Once that was established, Moore would call on the weakest student first. That had two effects: it stopped the course from turning into an uninterrupted series of lectures by the best student, and it made for a fierce competitive attitude in the class—nobody wanted to stay at the bottom. Moore encouraged competition. Do not read, do not collaborate-think, work by yourself, beat the other guy. Often a student who hadn't yet found the proof of Theorem 11 would leave the room while someone else was presenting the proof of it—each student wanted to be able to give Moore his private solution, found without any help. Once, the story goes, a student was passing an empty classroom, and, through the open door, happened to catch sight of a figure drawn on a blackboard. The figure gave him the idea for a proof that had eluded him till then. Instead of being happy, the student became upset and angry, and disqualified himself from presenting the proof. That would have been cheating—he had outside help!

Since, as I have already said, I have not really seen Moore in action in a serious mathematical class, I cannot guarantee that the description of the method I just offered is accurate in detail—but it is, I have been told, correct in spirit. I tried the method, experimented with it, kept running into tactical problems, worked out modifications that seemed to suit the students and the courses I was faced with—and I became a convert.

Some say that the only possible effect of the Moore method is to produce research mathematicians—but I don't agree. The Moore method is, I am convinced, the right way to teach anything and everything—it produces students who can understand and use what they have learned. It does, to be sure, instill the research attitude in the student—the attitude of questioning everything and wanting to learn answers actively but that's a good thing in every human endeavor, not only in mathematical research. There is an old Chinese proverb that I learned from Moore himself:

I hear, I forget; I see, I remember; I do, I understand.

One time at the University of Michigan I taught a beginning calculus class by something like the Moore method. It was a small class (about 15 or 16 students) and it was called an honors class. The designation did not mean that the students were geniuses, or even that they were mathematics majors: all it meant was that they were brighter than average, and, as it turned out, cooperative and pleasant to work with. The size of the class was right—at least for me; if it had been twice as large I don't think I could have handled it a la Moore. I have had Moore classes of 3 or 4 students and some of 18 or 20, and the latter, as far as I'm concerned, is the absolute maximum that the method works for. In one recent class of more than 40 I was able to get quite a bit of participation from the students and thus to avoid the straight sermon system—but the participation came from about half the class only. The other half sat stolidly no matter what I did—I didn't even learn the names of most of them before the term ended (or they mine?). Incidentally, too few students is also a disadvantage—2 is too few, and usually so is 3 or 4. It takes a critical mass to generate the kind of cooperative competition that makes the method work—the ideal class size is between 5 and 15.

I don't have a story of spectacular success to report about my honors calculus class. It went well enough, I am sure I didn't do them any harm, and at the end of the term they were at least as ready for the next course as I could have made them with any other method. To be sure, that was the first (and, I'm sorry to say, the only) time I tried Moore on a class at that level. I have faith that with more experience I could have made the method work for ordinary classes.

Another time I used the Moore method on an honors class in linear algebra with about 15 students. The first day of class I handed each student a set of 19 dittoed pages stapled together, and I told them that they now held the course in their hands. Those 19 pages contained the statements of fifty theorems, and nothing else. There were no definitions, there was no motivation, there were no explanations—nothing but fifty theorems, stated correctly but brutally, with no expository niceties. That, I told them, is the course. If you can understand, state, prove, exemplify, and apply those fifty theorems, you know the course, you know everything about linear algebra that this course is intended to teach you.

I will not, I told them, lecture to you, and I will not prove the theorems for you. I'll tell you, bit by bit as we go along, what the words mean, and I might from time to time indicate what this subject has to do with other parts of mathematics, but most of the classroom work will have to be done by you. I am challenging you to discover the proofs for yourselves, I am putting you on your honor not to look them up in a book or get outside help in any other way, and then I'll ask you to present in class the proofs you've discovered. The rest of you, the ones who are not doing the presenting, are supposed to stay on your toes mercilessly—make sure that the speaker gives a correct and complete proof, and demand from him whatever else is appropriate for understanding (such as examples and counterexamples).

They stared at me, bewildered and upset—perhaps even hostile. They had never heard of such a thing. They came here to learn something and now they didn't believe they would. They suspected that I was trying to get away with something, that I was trying to get out of the work I was paid to do. I told them about R. L. Moore, and they liked that, that was interesting. Then I gave them the basic definitions they needed to understand the statements of the first two or three theorems, and said "class dismissed".

It worked. At the second meeting of class I said, "O.K., Jones, let's see you prove Theorem 1", and I had to push and drag them along before they got off the ground. After a couple of weeks they were flying. They liked it, they learned from it, and they entered into the spirit of research—competition, discouragement, glory, and all.

If you are a teacher and a possible convert to the Moore method, don't make the mistake that my linear algebra students made: don't think that you'll do less work that way. It takes me a couple of months of hard work to prepare for a Moore course, to prepare the fifty theorems, or whatever takes their place. I have to chop the material into bite-sized pieces, I have to arrange it so that it becomes accessible, and I must visualize the course as a whole—what can I hope that they will have learned when it's over? As the course goes along, I must keep preparing for each meeting: to stay on top of what goes on in class, I myself must be able to prove everything. In class I must stay on my toes every second. I must not only be the moderator of what can easily turn into an unruly debate, but I must understand what is being presented, and when something fishy goes on I must interrupt with a firm but gentle "Would you explain that please?—I don't understand."

I am convinced that the Moore method is the best way to teach that there is—but if you try it, don't be surprised if it takes a lot out of you.

Moore and covering material

The two biggest obstacles to the success of the Moore method (or, for that matter, of teaching of any kind) are students who don't want to be there and students who want to be somewhere else. The two are not the same thing; let me explain.

By students who don't want to be there, I refer to required courses. If a student comes to me and asks my help to learn something that I already know, I am overjoyed. I am sure I can teach it to him, whatever it is, and I expect the process to be pleasant for both of us. If, however, he comes to me and says "I don't really want to know this stuff, but it's required that I get a C in it before I can go out and make a lot of money", then I'm unhappy.

Some universities used to require a mathematics course of all pre-med students (sometimes more than just one), for no reason except to serve as a sieve. The number of people preparing for admission to medical school was greater than the number that could be admitted—one way to lower the number of competitors was to flunk some in a mathematics course. Another reason for requirements is typified by chemists. They have to know some physics, and to learn the physics they have to learn some mathematics—it is a required course for reasons they don't fully understand and are inclined to resent. Similarly, students in the business school are required to learn some calculus, most of them for no reason at all and a few others for psychological reasons only: when, later in life, they meet an integral sign in an article on economics, they shouldn't be frightened out of their wits. A class of students who take a course because of such requirements is a sad and discouraging class. The highly motivated ones will work all right, and since most of them are not likely to be stupid, they will learn something, but the first prerequisite for the learning process to be both pleasant and effective—namely curiosity—is lacking, and that ruins it all. It ruins the teaching, it ruins the learning, it ruins the fun.

I dream of the ideal university, full of students who are full of intellectual curiosity. The subset of those among them who take a mathematics course do so because they want to know mathematics. They may be future doctors or chemists or executives in a shirt factory, but, for whatever reason, they *want* to find out what this mathematics stuff is about, and they come to me free willing and ask me to teach it to them. Oh, joy! If that really happened, I'd jump at the chance, and, at every chance I got, I'd use the Moore method. As matters stand, in this world, we teachers can only do the best we can with what we've got. Some of us do it in a good-natured fashion and some not, but we all teach a lot of mathematics as a trade. Under certain circumstances you remove parentheses, and here is how you do that; at other times you replace x by tan θ; and at still other times, very rarely, you choose δ to be $\varepsilon/3\sqrt{m}$ —follow the rules, and you won't go wrong. It's a living, and for those of us teachers who like mathematics, it leaves time to study the subject and think about it between classes, in the evening, and on weekends—it's not a bad life.

In any event, the Moore method is not certain to work with students who don't want to be there. Its second main obstacle (students who want to be somewhere else) is smaller than the first, but frightens some teachers anyway: how can you use the Moore method in Math. 307 and still be sure that your students will be prepared for Math. 407 next year? Isn't it true that you don't "cover" as much material that way?

Yes, it's true; no, it's not true. If the course description insists that you teach your 307 students each of a list of 31 theorems—teach in the sense of state, prove, and ask about on examinations—so that they'll be prepared to have the next 31 theorems taught to them in 407, then hurry up and start lecturing—it's better than getting fired. If, however, 307 is advanced calculus, say, and 407 is "baby real variables" (the story is the same for any other pair of courses in sequence), then ask yourself (and have your department ask itself) just what is the purpose of the game. Surely the course description is not an end in itself. The reason for the 31 theorems in 307 is to put the student in position to understand the theorems in 407, and the reason for them, in turn, is to understand some other theorems (for the mathematician) or to understand and apply them (for the chemist and the economist). To be able to understand and to be able to use—that's what we really try to teach, and that's what the Moore method does better than any other.

From my experience and from conversations with other users of the Moore method, orthodox and reformed, I conclude that the business of "covering" material is a red herring. Other methods cover more in less time, some say; Moore is a luxury we can't afford. I'm sure that's wrong. I feel strongly about it, strongly enough to interrupt the discussion of the Moore method, and say something about "covering" things.

There is no use trying to teach Calculus 2 to those who haven't learned Calculus 1—that seems to be true enough. Let me ask something, however, Moore or no Moore—is it true that the teacher of Calculus 1 always covers the six chapters that the course description demands and that, therefore, the students who passed Calculus 1 are ready to begin Chapter VII? Listen to almost any teacher of Calculus 2 at almost any university griping about his predecessor, the one who taught Calculus 1 last year: "Why, he didn't even mention partial fractions, and I don't know what he thinks he taught them about trigonometric substitution, but they sure don't know it. I had to spend three weeks doing first year stuff—no wonder I got behind." Description or no description, Moore or no Moore, the material that course N covers is never—hardly ever—what course $N + 1$ needs.

It's hard to teach a non-trivial course in probability to people who cannot use the exponential function. I taught one once, and the prerequisite was plainly stated (Calculus 2, or whatever it was called), and the prerequisite was supposed to have treated e^x. Five of my 25 students knew that it was its own derivative, and two of those had heard the rumor that it was essentially the only function with that property. Since, however, none of them had any facility with infinite series (and were uncomfortable even when I "reminded" them of the formula for the sum of a geometric series), the first two or three weeks of the probability course were in effect the last two or three weeks of Calculus 2. (Infinite series were in Chapter XI of the calculus text, and "nobody ever gets that far".)

That has nothing to do with the Moore method—it just seems to be a fact of life. What can be done about it? What seems to work is an intelligent syllabus for each course. An intelligent syllabus is one that tells you what tools you must have to be able to enter the course and cope with it, and what you should take with you when you leave it, so as to be able to cope with the next one. Both the start and the finish should be described explicitly. It is not good enough to refer to one book, the principal text, and say "the course consists of the first six chapters". Much better: "The indispensable prerequisites are the following ten theorems or techniques or formulas. (1) Integration by parts, as in Text A, pp. 81–93, or Text B, pp. 7–11, or Text C, pp. 207–234...." Similarly: "The minimum that the course is designed to cover is the following set of 31 theorems. (1) Term-by-term integration, as in ..., etc."

The syllabus should be explicit but not rigid. It should, to begin with, not try to fill every minute of the term; it should leave some time, say one third of the total, to the teacher's discretion. That is realistic, and it replaces the atmosphere of a cram session by something like the spirit of higher education. Every teacher has, and should be encouraged to have, private prejudices about what are the prettiest, or most useful, or most important parts of the course, and should be given the chance to insert those individualistic prejudices when they are appropriate. Teachers aren't, and shouldn't be, interchangeable parts of a machine, and students should have a chance to learn about different points of view. Another way that a syllabus could be rigid, and obviously shouldn't be, is to be thought of as something fixed for all time. Mathematics changes, fashions change, applications change, books change—syllabi should change as often as necessary. Every year is probably too often, and every ten years is probably not often enough—curriculum committees can easily stay on top of such matters.

Would syllabi have helped my probability course? Probably, in two ways. If a student sees the exponential function, and infinite series, in the minimum list that his calculus course should cover, he might act as part of his teacher's conscience. "When are we going to do infinite series? Is that really important? Will I need it when I take Statistics 3 next year?" He might, if the calculus teacher doesn't teach series, look it up and learn it by himself—improbable, in our modern university atmosphere, but not totally impossible. That's one of the two ways: the student is given a chance to act responsibly about the calculus syllabus. Emphasis: syllabi are not to be just secret score cards that the teacher knows about—they must be in the hands of the students. The other way a syllabus can help, and might have helped the probability course, is as a warning to the prospective student: if you haven't heard of exponentials and series, you're likely to be in trouble. Learn them first, or don't take probability.

A syllabus is a sheet of paper—it doesn't, by itself, work magic. One reason it could work, however, is that it spreads the responsibility, it helps the student and the teacher share it. It's no tragedy if not everything is covered, so long as everyone knows just what was supposed to be covered. Gaps can be filled, if we know what they are.

Many of my colleagues are pessimistic (or are they just realistic, or are they cynical?) about the likelihood that students will look up something in a book and learn by themselves. Students, many of us seem to think, are stupid and badly trained, and they must be spoon-fed. Sometimes that may be true, but so long as we believe it to be always true, we'll never change any of it. If business school students are too stupid to learn calculus except by spoon-feeding, and, even then, if what we teach them is just a bunch of mechanical rules, is it really wise to try to teach it to them? Some books are badly written and hard to read, and some students are stupid or badly trained but if we always act "realistically" and refuse to try to improve matters, we'll almost certainly go from bad to worse.

Syllabi, intelligently designed ones, would be a great help for advanced courses, graduate courses, as well as for elementary ones. Graduate students do come to us "free willing"—much more than most undergraduates—and they are somewhat less frightened of studying by themselves. Syllabi would help keep the courses straight (you better know what the closure of a set is before you sign up for the complex variable course), and they would be a great help with the part of the examination system that spans several courses. Most PhD programs have a qualifying examination as one of their major parts, and students frequently complain that they had algebra from Jones last year, but Robinson, who is teaching the course this year, emphasizes completely different topics, and he is the one who made up the exam. If both Jones and Robinson were bound (at least 2/3 bound) by a good syllabus, some of the grounds for the complaint might go away.

Syllabi help the teacher too, even the dilatory one, the one who finds himself two weeks from the end of term with six weeks of the syllabus still to go. You *can* present the last 12 theorems in two weeks instead of six, and you might even do it better than if you had six weeks to plough through all the computations. Here is how. For each theorem, describe the context it belongs to, the history that produced it, and the logical and psychological motivation that makes it interesting; state it, perhaps roughly, intuitively at first, and then precisely; and, finally use it—show what other results it makes contact with, and what follows from it. Take any of your favorite theorems and imagine yourself doing that—the fundamental theorem of projective geometry, or of Galois theory, or even of calculus, or, if you prefer, some results of a less sweeping kind, such as, say, the simplicity of the alternating groups, or the construction of the Cantor function. Your students will have to do without seeing you put your carefully prepared complicated proofs on the board. Will they really miss them very much? Proofs can be looked up. Contexts, histories, motivations, and applications are harder to find—that's what teachers are really for.

If you do what I just described, then, in part, you are already using a modified Moore method. It doesn't matter why you were behind the syllabus—because you told too many jokes in class, because you got mixed up in your earlier proofs, or because you cajoled the earlier proofs out of your students à la Moore. There is my point; the defense rests. What I just said is that with a little elasticity you can even adapt the Moore method to "covering" a previously fixed amount of material. You'll be covering a lot of it a lot better than your lecturing colleague (namely, the part that your students were led to discover and prove by themselves), and you'll be covering a part of it, all but the details of the proofs at the end, well enough so that your Moore-trained students are not likely to miss your doing for them what you've trained them to do for themselves.

How to be a pro

Anna, one of the mathematics department secretaries at Chicago, complained to me once when we met at a party and she had already consumed several gin and tonics. She was a good typist, and she didn't have much trouble picking up the art of technical, mathematical, typing—but she hated every minute of it. You have to look at every symbol separately, you have to keep changing "typits" or Selectric balls, you have to shift a half space up or down for exponents and subscripts, and you have no idea what you're doing. "I'm not going to spend my life flipping that damned carriage up and down", she said.

Another time Bruno, a well-known mathematician, asked me: "How did you manage to make your ten lectures at the CBMS conference all the same length? Isn't it true that some things take longer? When I did it, some of my talks were 45 minutes, and some 75."

More recently still I asked Calvin, a colleague: "Can you give a graduate student a ride to this month's Wabash functional analysis seminar?" "No", he said, "you better get someone else. I've been away, giving colloquium talks twice this month, and that's enough traveling."

I didn't make up any of this (except the names). Do you see what these three stories have in common? To me it's obvious, it jumps to the eye, and it horrifies me. What Anna, Bruno, and Calvin share and express is widespread and bad: it's the "me" attitude. It is the attitude that says "I do only what's important to me, and I am more important to me than the profession."

I think an automobile transmission mechanic should try to be the best automobile transmission mechanic he has the talent to be, and butlers, college presidents, shoe salesmen, and hod-carriers should aim for perfection in their professions. Try to rise, improve conditions if you can, and change professions if you must, but as long as you are a hodcarrier, keep carrying those hods. If you set out to be a mathematician, you must learn the profession, every part of it, and then work at it, profess it, live it as best you can. If you keep asking "what's there in it for me?", you're in the wrong business. If you're looking for comfort, money, fame, and glory, you probably won't get them, but if you keep trying to be a mathematician, you might.

An artist must know his medium and make the best possible use of it. If you set out to paint a miniature, don't complain that you don't have enough room. The size is an intrinsic part of the medium and you adapt to it the way you adapt to breathing on a mountain top—you have no choice. That's why Bruno's question sounded to me like asking why Michelangelo's statues aren't in the key of C or why Thackeray didn't write in cobalt blue. A vital part of the definition of each creative art is the medium in which it is expressed. An hour's lecture is, by its most common definition, a 50-minute performance, not a chapter in your book or a complete exposition of your recent work on harmonic analysis. When you agree to give a one-hour lecture, you do your best to give a one-hour lecture, making allowances for the level of the audience, estimating the probability of interruptions by questions, and organizing the material to fit the medium.

I didn't preach to Calvin, but if I had done so, my sermon would have said that you support an activity such as the Wabash seminar (I'll tell more about that seminar later) without even thinking about it—it's an integral part of professional life. You go to such seminars the way you get dressed in the morning, nod amiably to an acquaintance you pass on the street, or brush your teeth before you go to bed at night. Sometimes you feel like doing it and sometimes not, but you do it always—it's a part of life and you have no choice. You don't expect to get rewarded if you do and punished if you don't, you don't think about it—you just do it.

A professional must know every part of his profession and (we all hope) like it, and in the profession of mathematics, as in most others, there are many parts to know. To be a mathematician you have to know how to be a janitor, a secretary, a businessman, a conventioneer, an educational consultant, a visiting lecturer, and, last but not least, in fact above all, a scholar.

As a mathematician you will use blackboards, and you should know which ones are good and what is the best way and the best time to clean them. There are many kinds of chalk and erasers; would you just as soon make do with the worst? At some lecture halls you have no choice—you must use the overhead projector, and if you don't come prepared with transparencies, your audience will be in trouble. Word processors and typewriters, floppy disks and lift-off ribbons—ignorance is never preferable to bliss. Should you ditto your preprints, or use mimeograph, or multilith, or xerox? Who should make decisions about these trivialities—you, or someone who doesn't care about your stuff at all?

From time to time you'll be asked for advice. A manufacturer will consult you about the best shape for a beer bottle, a dean will consult you about the standing of his mathematics department, a publisher will consult you about the probable sales of a proposed textbook on fuzzy cohomology. Possibly they will be genteel and not even mention paying for your service; at other times they will refer delicately to an unspecified honorarium that will be forthcoming. Would they treat a surgeon the same way, or an attorney, or an architect? Would you? Could you?

I am sometimes tempted to tell people that I am a real doctor, not the medical kind; my education lasted a lot longer than their lawyer's and cost at least as much; my time and expertise are worth at least as much as their architect's. In fact, I do not use tough language, but I've long ago decided not to accept "honoraria" but to charge fees, carefully spelled out and agreed on in advance. I set my rates some years back, when I was being asked to review more text-books than I had time for. I'd tell the inquiring publisher that my fee is $1.00 per typewritten page or $50.00 per hour, whichever comes to less. Sometimes the answer was: "Oh, sorry, we didn't really want to spend that much", and at other times it was, matter-of-factly, "O.K., send us your bill along with your report". The result was that I had less of that sort of work to do and got paid more respectably for what I still did. My doctor, lawyer, and architect friends tell me that prices have changed since I established mine. The time has come, they say, to double the charges. The answers, they predict, will remain the same: half "no" and half "sure, of course". A professional mathematician talks about mathematics at lunch and at tea. The subject doesn't have to be hot new theorems and proofs (which can make for ulcers or ecstasy, depending) it can be a new teaching twist, a complaint about a fiendish piece of student skulldug-gery, or a rumor about an error in the proof of the four color theorem. At many good universities there is a long-standing tradition of brown-bag (or faculty club) mathematics lunches, and they are an important constituent of high qual-ity. They keep the members of the department in touch with one another; they make it easy for each to use the brains and the memory of all. At a few univer-sities I had a hand in establishing the lunch tradition, which then took root and flourished long after I left.

The pros go to colloquia, seminars, meetings, conferences, and international congresses—and they use the right word for each. The pros invite one another to give colloquium talks at their universities, and the visitor knows—should know—that his duty is not discharged by one lecture delivered between 4:10 and 5:00 p.m. on Thursday. The lunch, which gives some of the locals a chance to meet and have a relaxed conversation with the guest, the possible special-ists' seminar for the in-group at 2:00, the pre-lecture coffee hour at 3:00, and the post-lecture dinner and evening party are essential parts of the visitor's job description. It makes for an exhausting day, but that's how it goes, that's what it means to be a colloquium lecturer.

Sometimes you are not just a colloquium lecturer, but the "Class of 1909 Distinguished Visitor", invited to spend a whole week on the campus, give two or three mathematics talks and one "general" lecture, and, in between, mingle, consult, and interact. Some do it by arriving in time for a Monday afternoon talk, squeezing in a Tuesday colloquium at a sister university 110 miles down the road, reappearing for a Wednesday talk and a half of Thursday (spent mostly with the specialist crony who arranged the visit), and catching the plane home at 6:05 p.m. Bad show—misfeasance and nonfeasance—that's not what the idea is at all. When Bombieri came to Bloomington, I understood much of his first lecture, almost none of the second, and gave up on the third (answered my mail instead) but I got a lot out of his presence. I heard him hold forth on meromorphic functions at lunch on Monday, explain the Mordell conjecture over a cup of coffee Wednesday afternoon, and at dinner Friday evening guess at the probable standing of Euler and Gauss a hundred years from now. We also talked about clocks and children and sport and wine. I learned some mathematics and I got a little insight into what makes one of the outstanding mathematicians of our time tick. He earned his keep as Distinguished Visitor, not because he is a great mathematician (that helps!) but because he took the job seriously. It didn't take his talent to do that; that's something us lesser people can do too.

Mathematical talent is probably congenital, but aside from that the most important attribute of a genuine professional mathematician is scholarship. The scholar is always studying, always ready and eager to learn. The scholar knows the connections of his specialty with the subject as a whole; he knows not only the technical details of his specialty, but its history and its present standing; he knows about the others who are working on it and how far they have reached. He knows the literature, and he trusts nobody: he himself examines the original paper. He acquires firsthand knowledge not only of its intellectual content, but also of the date of the work, the spelling of the author's name, and the punctuation in the title; he insists on getting every detail of every reference absolutely straight. The scholar tries to be as broad as possible. No one can know all of mathematics, but the scholar can succeed in knowing the outline of it all: what are its parts and what are their places in the whole?

These are the things, some of the things, that go to make up a pro.

Paul R. Halmos was born in Hungary in 1916. He was graduated from the University of Illinois with a major in mathematics and philosophy in 1934 and earned his PhD degree in 1938. His first book, *Finite Dimensional Vector Spaces*, was published in 1942 and is still in print. He has written many other books, most recently *A Linear Algebra Problem Book* (1995) and *Logic as Algebra* (1998, with Steven Givant), both published by the Mathematical Association of America. His *I Have a Photographic Memory* (1987) wasn't really *written*, being a picture book, but it becomes more fascinating as the years go by. What a lot of neckties there were back then! He was justly famed for the clarity and elegance of his writing and lectures. He died in 2006.

To use resources optimally, it would be best if those teachers who adopt the Moore method are those who lecture poorly. When Halmos used it, his skill as an expositor was going to waste as he sat in a chair listening to students' more or less inept attempts to prove theorems. I was a member of such a class that he taught at the University of Michigan. Functional Analysis was its title, and its content was part of what later appeared in print as *A Hilbert Space Problem Book* (1967). We were given lists of problems (some of them could have been called theorems, but they were all presented as problems) and told to solve them and be prepared to present our solutions to the class. The problems were graded by stars, one to five, in order of estimated difficulty. I could knock off the one-star problems and almost all of the two-star ones, but considerably fewer than half of the three-star variety. I succeeded with perhaps one four-star problem.

It was always a pleasure when Paul would, before or after a student's presentation of a problem, make comments suggested by it. From Paul I would have preferred a non-Moore class.

12

Early Calculus

The first edition of the *Encyclopaedia Britannica* ("A Dictionary of Arts and Sciences Compiled upon a New Plan") was published in Edinburgh, 1768–1771. Its editor was William Smellie (pronounced "Smiley"), who may have written a good deal of it. It was a successful publication. When the time came for a second edition, its proprietors decided to include biographies, which caused Smellie to decline to have anything to do with it because biography is neither an art nor a science. The proprietors went ahead anyway and proceeded to get rich, while Smellie died in very reduced circumstances. The moral, I suppose, is that having principles can sometimes be costly.

There follows the *Encyclopaedia*'s article on fluxions, presenting all the reader would need to know about calculus in five large encyclopaedia pages. The article on chemistry, by contrast, consumed one hundred and fourteen pages in a three-volume work, which seems a disproportionate amount of space.

The five pages cover much of what in modern texts takes five hundred pages, or more. The rules for differentiation are all there, including the chain rule, and some techniques of integration, though "this branch of the art is imperfect" because some fluxions do not have simple fluents. Then there are applications: maxima and minima, points of inflection, areas, volumes, arc length, surface area—all there, and if the unknown author hadn't wasted quite a bit of space on a numerical example, even more could have been included.

Some space was saved by not considering trigonometric functions, or exponentials and logarithms. The way around that is to use infinite series.

The text has several obvious misprints, and there seems to be a figure missing, "the curve ARS" referred to in Problem 3 never being pictured. I found the article interesting reading. It appears essentially as printed, though I have replaced overlines as symbols of grouping with parentheses: $a \times (b+c)$ instead of $a \times \overline{b+c}$.

FLUXIONS, a method of calculation which greatly facilitates computations in the higher parts of mathematics. Sir Isaac Newton and Mr Leibnitz contended for the honour of inventing it. It is probable they had both made progress in the same discovery, unknown to each other, before there was any publication on the subject.

In this branch of mathematics magnitudes of every kind are supposed generated by motion: a line by the motion of a point, a surface by the motion of a line, and solid by the motion of a surface. And some part of a figure is supposed generated by an uniform motion; in consequence of which the other parts may increase uniformly or with an accelerated or retarded motion, or may decrease in any of these ways; and the computations are made by tracing the comparative velocities with which the parts flow.

Fig. 1. If the parallelogram *ABCD* be generated by uniform motion of the line *AB* toward *CD* while it moves from *FE* towards *fe*, while the line *BF* receives the increment *Ff*, and the figure will be increased by the parallelogram *Fe*; the line *FE* in this case undergoes no variation.

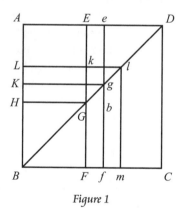

Figure 1

The fluxion of any magnitude at any point is the increment that it would receive in any given time, supposing it to increase uniformly from that point; and as the measures will be the same, whatever the time be, we are at liberty to suppose it less than any assigned time.

The first letters in the alphabet are used to represent invariable quantities; the letters x, y, z variable quantities; and the same letters with points over them $\dot{x}, \dot{y}, \dot{z}$ represent their fluxions.

Therefore if $AB = a$, and $EF = x$; Ff, the fluxion of BF, will be $= \dot{x}$, and Fe, the flexion of AF, $= a\dot{x}$.

If the rectangle be supposed generated by the uniform motion of FG towards CD, at the same time HG moves uniformly towards AD, the point G keeping always on the diagonal, the lines FG HG will flow uniformly; for while Bf receives the increment Ff and HB, the increment HK, FG will receive the increment bg and HG the increment bg, and they will receive equal increments in equal successive times. But the parallelogram will flow with an accelerated motion; for while F flows to f and H to K, it is increased by the gnomon KGf; but while F and H flow through the equal spaces $fmKL$ it is increased by the gnomon Lgm greater than KGf; consequently when fluxions of the sides of a parallelogram are uniform, the fluxion of the parallelogram increases continually.

The fluxion of the parallelogram $BHGF$ is the two parallelograms KG and Gf for though the parameter receives an increment of the gnomon KGf, while its sides flow to f and K, the part gG is owing to the additional velocity wherewith the parallelogram flows during that time; and therefore is no part of the measure of the fluxion, which must be computed by supposing the parameter to flow uniformly as it did at the beginning, without any acceleration.

Therefore if the sides of a parallelogram be x and y, their fluxions will be $\dot{x}\dot{y}$; and the fluxion of the parallelogram $x\dot{y} + y\dot{x}$; and if $x = y$, that is, if the figure be a square, the fluxion of x^2 will be $2x\dot{x}$.

Fig. 2. Let the triangle ABC be described by the uniform motion of DE from A towards B, the point E moving in the line DF, so as always to touch the lines AC, CB; while D moves from A to F, DE is uniformly increased, and the increase of the triangle is uniformly accelerated. When DE is in the position FC, it is a maximum. As D moves from F to B, the line FC decreases, and the triangle increases, but with a motion uniformly retarded.

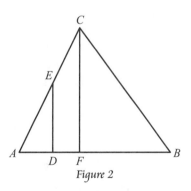

Figure 2

Fig. 3. If the semicircle AFB be generated by the uniform motion of CD from A towards B. while C moves from A to G, the line CD will increase, but with a retarded motion; the circumference also increases with a retarded motion, and the circular space increases with an accelerated motion, but not uniformly, the degrees of acceleration growing less as CD approaches to the

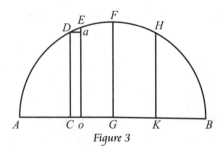

Figure 3

portion *GF*. When *C* moves from *G* to *B*, it decreases with a motion continually accelerated, the circumference increases with a motion continually accelerated, and the area increases with a motion continually retarded, and more quickly retarded as *CD* approaches to *B*.

The fluxion of a quantity which decreases is to be considered as negative.

When a quantity does not flow uniformly, its fluxion may be represented by a variable quantity, or a line of a variable length; the fluxion of such a line is called the second fluxion of the quantity whole fluxion that line is and if it be variable, a third fluxion may be deduced from it, and higher orders from there in the same manner: the second fluxion is represented by two points, as $\ddot{x}$.

The increment a quantity receives by flowing for any given time, contains measures of all the different orders of fluxions; for if it increases uniformly, the whole increment is the first fluxion; and it has no second fluxion. If it increases with a motion uniformly accelerated, the part of the increment occasioned by the first motion measures the first fluxion, and the part occasioned by the acceleration measures the second fluxion. If the motion be not only accelerated, but the degree of acceleration continually increased, the two first fluxions are measured as before; and the part of the increment occasioned by the additional degree of acceleration measures the third; and so on. These measures require to be corrected, and are only mentioned here to illustrate the subject.

Direct Method

Any flowing quantity being given, to find its fluxion.

Rule I. To find the fluxion of any power of a quantity, multiply the fluxion of the root by the exponent of the power, and the product by a power of the same root less by unity than the given exponent.

The fluxion of x^3 is $3x^2\dot{x}$, of x^n $nx^{n-1}\dot{x}$; for the root of x^n is x, whole fluxion is $\dot{x}$; which multiplied by the exponent n, and by a power of x less by unity than n, gives the above fluxion.

If x receive the increment $\dot{x}$, it becomes $x + \dot{x}$; raise both to the power of n, and x^n becomes

$$x^n + nx^{n-1}\dot{x} + \frac{n(n-1)}{2}x^{n-2}\dot{x}^2 + , \&c;$$

but all the parts of the increment, except the first term, are owing to the accelerated increase of x^n, and form measures of the higher fluxions. The first term only measures the first fluxion; the fluxion of $(a^2 + z^2)^{3/2}$ is $\frac{3}{2} \times 2z\dot{z} \times (a^2 + z^2)^{1/2}$; for put $x = a^2 + z^2$, we have $\dot{x} = 2z\dot{z}$, and the fluxion of $x^{3/2}$, which is equal to the proposed fluent, is $\frac{3}{2}x^{1/2}\dot{x}$, for which substituting the values of x and $\dot{x}$, we have the above fluxion.

RULE II. To find the fluxion of the product of several variable quantities multiplied together, multiply the fluxion of each by the product of the rest of the quantities, and the sum of the products thus arising will be the fluxion sought.

Thus the fluxion of xy, is $x\dot{y} + y\dot{x}$; that of xyz, is $xy\dot{z} + xz\dot{y} + yz\dot{x}$, and that of $xyzu$, is $xyz\dot{u} + zyu\dot{z} + xzu\dot{y} + yzu\dot{x}$.

RULE III. To find the fluxion of a fraction.—From the fluxion of the numerator multiplied by the denominator, subtract the fluxion of the denominator multiplied by the numerator, and divide the remainder by the square of the denominator.

Thus, the fluxion of $\dfrac{x}{y}$ is $\dfrac{y\dot{x} - x\dot{y}}{y^2}$; that of $\dfrac{x}{x+y}$, is

$$\frac{\dot{x} \times (x+y) - (\dot{x} + \dot{y}) \times x}{(x+y)^2} = \frac{y\dot{x} - x\dot{y}}{(x+y)^2}.$$

RULE IV. In complex cases, let the particulars be collected from the simple rules and combined together.

The fluxion of

$$\frac{x^2 y^2}{z} \text{ is } \frac{(2x^2 y\dot{y} + 2y^2 x\dot{x}) \times z - x^2 y^2 \dot{z}}{z^2};$$

for the fluxion of x^2 is $2x\dot{x}$, and of y^2 is $2y\dot{y}$, by Rule I, and therefore the fluxion of $x^2 y^2$ (by Rule II) $2x^2 y\dot{y} + 2y^2 x\dot{x}$; from which, multiplied by z, (by Rule III) and subtracting from it the fluxion of the denominator z, multiplied by the numerator, and dividing the whole by the square of the denominator, gives the above fluxion.

RULE IV. The second fluxion is derived from the first, in the same manner as the first from the flowing quantity.

Thus the fluxion of x^3, $3x^2\dot{x}$; its second, $6x\dot{x}^2\,3x^2\ddot{x}$ (by Rule II) ; and so on: but if $\dot{x}$ be invariable, $\ddot{x} = 0$, and the second fluxion of $x^3 = 6x\dot{x}^2$.

PROB. 1. *To determine maxima and minima.*

When a quantity increases, its fluxion is positive; when it decreases, it is negative; therefore when it is just betwixt increasing and decreasing, its fluxion is = 0.

RULE. Find the fluxion, make it = 0, whence an equation will result that will give an answer to the question.

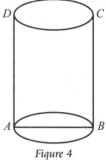

Fig. 4. EXAMP. To determine the dimensions of a cylindric measure *ABCD*, open at the top, which shall contain a given quantity (of liquor, grain, &c.) under the least internal superficies possible.

Let the diameter $AB = x$, and the altitude $AD = y$; moreover, let p (3.14159, &c.) denote the periphery of the circle whole diameter is unity, and let c be the given content of the cylinder. Then it will be $1 : p ::$ $x : (px)$ the circumference of the base; which, mul-

Figure 4

tiplied by the altitude y, gives pxy for the concave superficies of the cylinder. In like manner, the area of the base, by multiplying the same expression into 1/4 of the diameter x, will be found $= px^2/4$; which drawn into the altitude y, gives $px^2y/4$ for the solid content of the cylinder; which being made $= c$, the concave surface pxy will be found $= 4c/x$, and consequently the whole surface $= \dfrac{4c}{x} + \dfrac{px^2}{4}$: Whereof the fluxion, which is $-\dfrac{4c\dot{x}}{x^2} + \dfrac{px\dot{x}}{2}$ being put = 0, we shall get $-8c + px^3 = 0$; and therefore $x = 2\sqrt[3]{c/p}$: further, because $px^3 = 8c$, and $px^2y = 4c$, it follows, that $x = 2y$; whence y is also known, and from which it appears, that the diameter of the base must be just the double of the altitude.

Fig. 7. To find the longest and shortest ordinates of any curve, *DEF*, whose equation or the relation which the ordinates bear to the abscissas is known.

Make *AC* the abscissa x, and *CE* the ordinate $= y$; take a value y in terms of x, and find its fluxion: which making = 0, an equation will result whose roots give the value of x when y is a maximum or minimum.

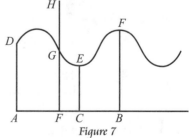

Figure 7

To determine when it is a maximum and when a minimum, take the value of *y*, when *x* is a little more than the root of the equation so found, and it may be perceived whether it increases or decreases.

If the equation has an even number of equal roots, *y* will be neither maximum nor minimum when its fluxion is = 0.

PROB. 2. *To draw a tangent to any curve.*

Fig. 5. When the abscissa *CS* of a curve moves uniformly from *A* to *B*, the motion of the curve will be retarded if it be concave, and accelerated if convex towards *AB*; for a straight line *TC* is described by an uniform motion, and the fluxion of the curve at any point is the same as the fluxion of the tangent, because

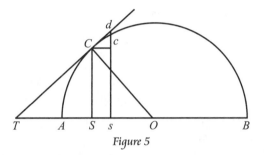

Figure 5

it would describe the tangent if it continued to move equally from that point. Now if *Ss* or *Cc* be the a fluxion of the base, *Cd* will be the fluxion of the tangent, and *dc* of the ordinate. And because the triangles *TSC*, *Ccd*, are equiangular, *dc* : *cC* :: *CS* : *ST*, wherefore

RULE. Find a fourth proportional to the fluxion of the ordinate valued in terms of the abscissa, the fluxion of the abscissa, and the ordinate, and it determines the line *ST*, which is called the semi-tangent, and *TC* joined is a tangent to the curve.

Fig. 6. EXAMP. To draw a right line *CT*, to touch a given circle *BCA* in a given point *C*.

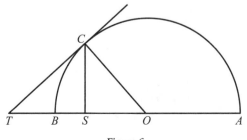

Figure 6

Let *CS* be perpendicular to the diameter *AB*, and put $AB = a$, $BS = x$, and $SC = y$: Then, by the property of the circle, $y^2(CS^2) = BS \times AS$ $(= x \times (a - x)) = ax - x^2$; whereof the fluxion being taken, in order to determine the ratio of $\dot{x}$ and $\dot{y}$, we get $2y\dot{y} = a\dot{x} - 2x\dot{x}$; consequently

$$\frac{\dot{x}}{\dot{y}} = \frac{2y}{a - 2x} = \frac{y}{\frac{1}{2}a - x};$$

which multiplied by *y*, gives

$$\frac{yx}{y} = \frac{y^2}{\frac{1}{2}a - x} = \text{ the subtangent } ST.$$

Whence (*O* being supposed the center) we have

$$OS\left(\tfrac{1}{2}a - x\right) : CS(y) :: CS(y) : ST;$$

which we also know from other principles.

PROB 3. *To determine points of contrary flexure in curves.*

Fig. 7. Supposing *C* to move uniformly from *A* to *B*, the curve *DEF* will be convex towards *AB* when the celerity of *E* increases, and concave when it decreases; therefore at the point where it ceases to be convex and begins to be concave, or the opposite way, the celerity of *E* will be uniform, that is, *CD* will have no second fluxion. Therefore,

RULE. Find the second fluxion of the ordinate in terms of the abscissa, and make it = 0; and from the equation that arises you get a value of the abscissa, which determines the point of contrary flexure.

EX.   Let the nature of the curve *ARS* be defined by the equation $ay = a^{1/2}x^{1/2} + xx$, (the abscissa *AF* and the ordinate *FG* being, as usual, represented by *x* and *y* respectively). Then $\dot{y}$, expressing the celerity of the point *r*, is the line *FH*, will be equal to

$$\frac{\tfrac{1}{2}a^{1/2}x^{-1/2}\dot{x} + 2x\dot{x}}{a} :$$

Whose fluxion, or that of $\tfrac{1}{2}a^{1/2}x^{-1/2} + 2x$ (because *a* and $\dot{x}$ are constant) must be equal to nothing; that is, $\tfrac{1}{4}a^{3/2}x^{-3/2}\dot{x} + 2\dot{x} = 0$: Whence $a^{3/2}x^{-3/2} = 8$, $a^{3/2} = 8x^{3/2}$, $64x^3 = a^3$, and $x = \tfrac{1}{4}a = AF$; therefore

$$FG\left(\frac{a^{3/2}x^{1/2} + xx}{a}\right) = \frac{9}{16}a :$$

From which the position of the point *G* is given.

PROB. 4. *To find the radii of curvature.*

The curvature of a circle is uniform in every point, that of every other curve continually varying: and it is measured at any point by that of a circle whore radius is of such a length as to coincide with it in curvature in that point.

All curves that have the same tangent have the same first fluxion, because the fluxion of a curve and its tangent are the same. If it moved uniformly on from the point of contact, it would describe the tangent. And the deflection from the tangent is owing to the acceleration or retardation of its motion, which is measured by its second fluxion; and consequently two curves which have not only the same tangent, but the same curvature at the point of contact, will have both their first and second fluxions equal. It is easily proven from thence, that the radius of curvature is $= \dfrac{\dot{z}^3}{-\dot{x}\ddot{y}}$ where x, y, and z represent the abscissa, ordinate, and curve respectively.

EXAMP. Let the given curve be the common parabola, whose equation is $y = a^{1/2}x^{1/2}$: then will

$$\dot{y} = \frac{1}{2}a^{1/2}\dot{x}x^{-1/2} = \frac{a^{1/2}\dot{x}}{2x^{1/2}},$$

and (making $\dot{x}$ constant)

$$\ddot{y} = -\frac{1}{2}\times\frac{1}{2}a^{1/2}\dot{x}^2 x^{-3/2} = \frac{-a^{1/2}\dot{x}^2}{4x^{3/2}}:$$

Whence

$$\dot{x}\left(\sqrt{\dot{x}^2 + \dot{y}^2}\right) = \frac{\dot{x}}{2}\sqrt{\frac{4x+a}{x}},$$

and the radius of curvature

$$\left(\frac{\dot{z}^3}{-\dot{x}\ddot{y}}\right) = \frac{(a+4x)^{3/2}}{2\sqrt{a}}:$$

Which at the vertex, where $x = 0$, will be $= \frac{1}{2}a$.

INVERSE METHOD.

From a given fluxion to find a fluent.

This is done by tracing back the steps of the direct method. The fluxion of x is $\dot{x}$; and therefore the fluent of $\dot{x}$ is x: but as there is no direct method of finding fluents, this branch of the art is imperfect. We can assign the fluxion of every fluent, but we cannot assign the fluent of a fluxion, unless it be such a one as may be produced by some rule in the direct method from a known fluent.

GENERAL RULE. Divide by the fluxion of the root, add unity to the exponent of the power, and divide by the exponent so increased.

For, dividing the fluxion $nx^{n-1}\dot{x}$ by $\dot{x}$ (the fluxion of the root x) it becomes nx^{n-1}; and, adding 1 to the exponent $(n-1)$ we have nx^n; which, divided by n, gives x^n, the true fluent of $nx^{n-1}\dot{x}$.

Hence (by the same rule) the

Fluent of $3x^2\dot{x}$ will be $= x^3$;

That of $8x^2\dot{x} = \dfrac{8x^3}{3}$;

That of $2x^5\dot{x} = \dfrac{x^6}{3}$;

That of $y^{1/2}\dot{y} = \dfrac{2}{3}y^{3/2}$.

Sometimes the fluent so found requires to be corrected. The fluxion of x is $\dot{x}$ and the fluxion of $a + x$ is also $\dot{x}$ because a is invariable, and has therefore no fluxion.

Now when the fluent of $\dot{x}$ is required, it must be determined, from the nature of the problem, whether any invariable part, as a, must be added to the variable part x.

When fluents cannot be exactly found, they can approximated by infinite series.

EX. Let it be required to approximate the fluent of

$$\frac{(a^2 - x^2)^{1/2}}{(c^2 - x^2)^{1/2}} \times x^n \dot{x}$$

in an infinite series. The value of

$$\frac{(a^2 - x^2)^{1/2}}{(c^2 - x^2)^{1/2}},$$

expressed in a series is

$$\frac{a}{c} + \left(\frac{a}{2c^3} - \frac{1}{2ac}\right) \times x^2 + \left(\frac{3a}{8c^5} - \frac{1}{4ac^3} - \frac{1}{8a^3c}\right)$$

$$\times x^4 + \left(\frac{5a}{16c^7} - \frac{3}{16ac} - \frac{1}{16a^3c^3} - \frac{1}{16a^5c}\right) \times x^6 + \&\text{c.}$$

Which value being therefore multiplied by $x^n\dot{x}$, and the fluent taken (by the common method) we get

$$\frac{ax^{n+1}}{(n+1)\times c}+\left(\frac{a}{2c^3}-\frac{1}{2ac}\right)\times\frac{x^{n+3}}{n+3}+\left(\frac{3a}{8c^5}-\frac{1}{4ac^3}-\frac{1}{8a^3c}\right)$$

$$\times\frac{x^{n+5}}{n+5}+\left(\frac{5a}{16c^7}-\frac{3}{16ac^5}-\frac{1}{16a^3c^3}-\frac{1}{16a^5c}\right)\times\frac{x^{n+7}}{n+7}+\ \&c.$$

PROB. 1. *To find the area of any curve.*

RULE. Multiply the ordinate by the fluxion of the abscissa, and the product gives the fluxion of the figure, whose fluent is the area of the figure.

EXAMP. 1. Fig. 8. Let the curve *ARMH*, whose area you will find, be the common parabola. Let *u* represent the area, and $\dot{u}$ its fluxion.

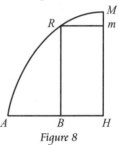
Figure 8

In which case the relation of *AB* (*x*) and *BR* (*y*) being expressed by $y^2 = ax$ (where *a* is the parameter) we thence get $y = a^{1/2}x^{1/2}$; and therefore

$$\dot{u} = RmHB\ (= yx) = a^{1/2}x^{1/2}\dot{x};$$

whence

$$u = \tfrac{2}{3}\times a^{1/2}x^{3/2} = \tfrac{2}{3}a^{1/2}x^{1/2}\times x = \tfrac{2}{3}yx$$

(because $a^{1/2}x^{1/2} = y$): hence a parabola is 2/3 of a rectangle of the same base and altitude.

EXAMP. 2. Let the proposed curve *CSDR* (fig. 9) be of such a nature, that (supposing *AB* unity) the sum of the areas *CSTBC* and *CDGBC* answering to any two proposed abscissas *AT* and *AG*, shall be equal to the area *CRNBC*, whose corresponding abscissa *AN* is equal to *AT* · *AG*, the product of the measures of the two former abscissas.

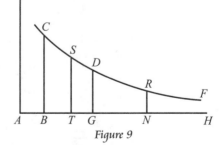
Figure 9

First, in order to determine the equation of the curve, (which must be known before the area can be found) let the ordinates *GD* and *NR* move parallel to themselves towards *HF*; and then having put *GD* = *y*, *NR* = *z*, *AT* = *a*, *AG* = *s*, and *AN* = *u*, the fluxion of the area *CDGB* will be represented by $y\dot{s}$, and that of the area *CRNB* by $z\dot{u}$: which two expressions must, by the nature of the problem, be equal to each other; because the latter area *CRNB* exceeds the former *CDGB* by the area *CSTB*, which is here considered as a constant quantity: and it is evident, that two expressions, that differ only by a constant quantity, must always have equal fluxions.

Since, therefore, $y\dot{s}$ is $=z\dot{u}$ and $u = as$, by hypothesis, it follows, that $\dot{u} = a\dot{s}$, and that the first equation (by substituting for $\dot{u}$) will become $y\dot{s} = az\dot{s}$, or $y = az$, or lastly $ys = zas$, that is, $GD \cdot AG = NR \cdot AN$: therefore, GD : NR :: $AN : AG$; whence it appears, that every ordinate of the curve is reciprocally as its corresponding abscissa.

Now, to find the area of the curve so determined, put $AB = 1$, $BC = b$, and $BG = x$: then, since AG $(1 + x)$: AB (1) :: BC (b) : GD (y) we have $y = \dfrac{b}{1+x}$, and consequently

$$\dot{u} \ (= y\dot{x}) = \frac{b\dot{x}}{1+x} = b \times (\dot{x} - x\dot{x} + x^2\dot{x} - x^3\dot{x} + x^4\dot{x} - \ \&\text{c.}).$$

Whence, *BGDC*, the area itself will be

$$= b \times \left(x - \frac{x^2}{2} + \frac{x^3}{3} - \frac{x^4}{4} + \frac{x^5}{5}, \ \&\text{c.} \right)$$

which was to be found.

Hence it appears, that as these areas have the same properties as logarithms, this series gives an easy method of computing logarithms; and the fluent may be found by means of a table of logarithms, without the trouble of an infinite series: and every fluxion whose fluent agrees with any known logarithmic expression, may be found the same way. Hence the fluents of fluxions of the following forms are deduced.

The fluent of $\quad \dfrac{\dot{x}}{\sqrt{x^2 \pm a^2}} = $ hyp. log. of $x + \sqrt{x^2 \pm a^2}$;

of $\quad \dfrac{\dot{x}}{\sqrt{2ax + xx}} = $ hyp. log. $a \times x + \sqrt{2ax + x^2}$;

of $\quad \dfrac{2a\dot{x}}{a^2 - x^2} = $ hyp. log. of $\dfrac{a+x}{a-x}$;

and of $\quad \dfrac{2a\dot{x}}{x\sqrt{a^2 \pm x^2}} = $ hyp. log. $\dfrac{a - \sqrt{a^2 \pm x^2}}{a\sqrt{a^2 \pm x^2}}$.

PROB. 2. *To determine the length of curves.*

Fig. 5. Because Cdc is a right-angled triangle, $Cd^2 = Cc^2 + dc^2$; wherefore the fluxions of the abscissa and ordinate being taken in the same terms and squared, their sum gives the square of the fluxion of the curve; whose root being extracted, and the fluent taken, gives the length of the curve.

EXAMP. To find the length of a circle from its tangent. Make the radius AO (fig. 5) $= a$, the tangent of $AC = t$, and its secant $= s$, the curve $= z$, and its fluxion $= \dot{z}$; because the triangles OTC, OCS, are similar, $OT : OC :: OC : OS$; whence

$$OS = \frac{a^2}{s}, \quad \text{and} \quad SA = a - \frac{a^2}{s} = a - \frac{a^2}{\sqrt{a^2 + t^2}} :$$

whose fluxion is

$$\frac{a^2 t \dot{\imath}}{(a^2 + t^2)^{3/2}} ;$$

and because the triangles OTC, dCc are similar,

$$TC (= t) : TO \left(= \sqrt{a^2 + t^2}\right) :: Cc \left(= \frac{a^2 t \dot{\imath}}{(a^2 + t^2)^{3/2}}\right) : Cd = \frac{a^2 t}{a^2 + t^2} =$$

fluxion of the curve. Now by converting this into an infinite series, we have the fluxion of the curve

$$= \dot{\imath} - \frac{t^2 \dot{\imath}}{a^2} + \frac{t^4 \dot{\imath}}{a^4} - \frac{t^6 \dot{\imath}}{a^6}, \text{ \&c.}$$

and consequently

$$z = t - \frac{t^3}{3a^2} + \frac{t^5}{5a^4} - \frac{t^7}{7a^6} + \frac{t^9}{9a^8}, \text{ \&c.} = AR.$$

Where, if (for example's sake) AR be supposed an arch of 30 degrees, and AO (to render the operation more easy) be put $=$ unity, we shall have $t = \sqrt{1/3} = .5773502$ (because $Ob\sqrt{1/4} : bR(1/2) :: OA(1) : AT(t) = \sqrt{1/3}$). Whence,

$$t^3 \left(= t \times t^2 = t \times \frac{1}{3}\right) = .1924500$$

$$t^5 \left(= t^3 \times t^2 = \frac{t^3}{3}\right) = .0641500$$

$$t^7 \left(= t^5 \times t^2 = \frac{t^5}{3}\right) = .0213833$$

$$t^9 \left(= t^7 \times t^2 = \frac{t^7}{3}\right) = .0071277$$

$$t^{11} \left(= t^9 \times t^2 = \frac{t^9}{3}\right) = .0023759$$

$$t^{13} \left(= t^{11} \times t^2 = \frac{t^{11}}{3}\right) = .0007919$$

$$t^{15} \left(= t^{13} \times t^2 = \frac{t^{13}}{3}\right) = .0002639$$

And therefore

$$AR = .5773502 - \frac{.1934500}{3} + \frac{.0641500}{5} - \frac{.0213833}{7} + \frac{.0071277}{9} - \frac{.0023759}{11}$$
$$+ \frac{.0007919}{13} - \frac{.0002639}{15} + \frac{.0000879}{17} - \frac{.0000293}{19} + \frac{.0000097}{21} - \frac{.0000032}{23}$$
$$= .5235987 :$$

for the length of an arch of 30 degrees, which multiplied by 6 gives 3.141592 + for the length of the semi-periphery of the circle whose radius is unity.

Other series may be deduced from the versed line, sine and secant; and these are of use for finding fluents which cannot be expressed in finite terms. For,

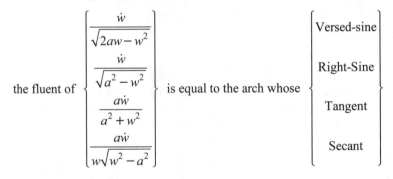

$$\text{the fluent of } \begin{Bmatrix} \dfrac{\dot{w}}{\sqrt{2aw - w^2}} \\[2ex] \dfrac{\dot{w}}{\sqrt{a^2 - w^2}} \\[2ex] \dfrac{a\dot{w}}{a^2 + w^2} \\[2ex] \dfrac{a\dot{w}}{w\sqrt{w^2 - a^2}} \end{Bmatrix} \text{ is equal to the arch whose } \begin{Bmatrix} \text{Versed-sine} \\[2ex] \text{Right-Sine} \\[2ex] \text{Tangent} \\[2ex] \text{Secant} \end{Bmatrix}$$

is $\frac{w}{a}$ and Radius Unity.

PROB. 3. *To find the contents of a solid.*

Let the surface of the generating plane be multiplied by the space it passes through in any time, the product will give a solid which is the fluxion of the solid required: the surface must therefore be computed in terms of x, which represents the line or axis on which it moves, and by its motion on which the fluxion is to be measured, and the fluent found will give the contents of the solid.

EXAMP. Let it be proposed to find the content of a cone *ABC*, fig. 10.

Put the given altitude (*AD*) of the cone = *a*, and let the semi-diameter (*BD*) of its base = *b*, the solid = *s*, its fluxion = *ṡ*, and the area of a circle, whose radius is unity = *p*: then the distance (*AF*) of the circle *EG*, from the vertex *A*, being denoted by *x*, &c. we have, by similar triangles, as

$$a:b::x:EF(y) = \frac{bx}{a}.$$

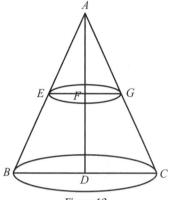

Whence in this case,

$$\dot{s} \left(= p\dot{y}^2\dot{x}\right) = \frac{pb^2x^2\dot{x}}{a^2};$$

and consequently

$$s = \frac{pb^2x^3}{3a^2};$$

which, when *x* = *a* (= *AD*) gives

$$\frac{pb^2a}{3} = \left(p \times BD^2 \times \frac{1}{3}AD\right)$$

for the content of the whole cone *ABC*: which appears from hence to be just 1/3 of a cylinder of the same base and altitude.

PROB. 4. *To compute the surface of any solid body.*

The fluxion of the surface of the solid is equal to the periphery of the surface, by whole motion the solid is generated, multiplied by its velocity on the edge of the solid, and the computation is made as in the foregoing.

EXAMP. Fig. 11. Let it be proposed to determine the convex superficies of a cone *ABC*.

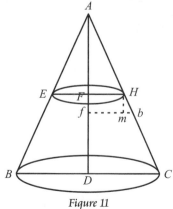

Figure 10

Figure 11

Then, the semi-diameter of the base (BD, or CD) being put $= b$, the slanting line, or hypotenuse $AC = c$, and FH (parallel to DC) $= y$, $AG = z$; the surface $= w$, its fluxion $= \dot{w}$, and $p =$ the periphery of a circle whole diameter is unity, we shall, from the similarity of the triangles ADC and Hmb, have

$$b : c :: \dot{y}(mh) : \dot{z}(Hh) = \frac{cy}{b} : \text{ whence } \dot{w}(2py\dot{z}) = \frac{2pcy\dot{y}}{b} ;$$

and consequently

$$w = \frac{pcy^2}{b}.$$

This, when $y = b$, becomes $= pcb = p \cdot DC \cdot AC =$ the convex superficies of the whole cone ABC: which therefore is equal to a rectangle under half the circumference of the base and the slanting line.

The method of fluxions is also applied to find the centres of gravities, and oscillation of different bodies; to determine the paths described by projectiles and bodies acted on by central forces, with the laws of centrepetal force in different curves; the retardates given to motions performed in resisting media; the attractions of bodies under different forms; the direction of wind, which has the greatest effect on an engine; and to solve many other curious and useful problems.

As the article shows, it is possible to do calculus using the Newtonian notation, but it has no advantage at all over the notation we now use. It was not until the first half of the nineteenth century that the "d-ism of Leibniz" finally replaced the "dot-age of Newton" in England.

The current edition of the *Brittanica* has an improved treatment of calculus.

13
Problems

Problems are the heart of mathematics. Trying to solve problems, from finding the digits that the letters stand for in

S E N D

M O R E

M O N E Y

to proving that all the non-trivial zeros of the Riemann zeta function lie on the line $x = \frac{1}{2}$, accounts for a large proportion of mathematical activity. It is no accident that four journals of the Mathematical Association of America contain sections of problems and solutions, nor that its list of best sellers always contains problem books. The appeal of problems is widespread and constant.

Problems appear in the earliest surviving mathematical writings. The Rhind papyrus (c. 1650 BC) is little more than a collection of problems and solutions. There are Babylonian clay tablets from about the same period with mathematics on them, and they too seem to be aids for solving problems.

Besides being fun to do (especially if you succeed in solving one now and then), problems are necessary for teaching. That is how mathematics is learned, and how it has always been learned: students read or hear about a mathematical idea and then do problems so as to make it their own. The ancient Egyptians were a very practical people—there are no ancient Egyptian papyri about philosophy—but the Rhind papyrus contains some very non-practical problems. For example, number 64 asks for an arithmetic progression with ten terms whose sum is 10 with difference 1/8. Though phrased as a problem in dividing grain, no one in ancient Egypt or since has had to divide grain that way. Students were required to do the problem so that they could master the mathematical ideas in order to solve more practical problems.

Ideas for good problems are not plentiful, so when one if found it stays around, being repeated (without credit) by author after author. If the person who first thought of the problem of fencing in the maximum area next to a river with a given amount of fence had a dollar for every time someone has had to solve it, he or she would be able to buy a large acreage, and have money left over for all the fencing needed to enclose it.

If new ideas aren't readily available, we can always extend and generalize old ones. So, in the fencing problem the shape of the field can change, or north-south fencing can be more expensive than that going east and west, and the fertile minds of mathematicians have generated another problem to work on. Sometimes the problems can become even further-fetched than the originals.

Of course there are fashions in problems as there are in clothes, and a historian of problems could spend many years finding them. Unfortunately, the supply of historians of mathematical problems is very small, probably because there is no demand for them. Stanley Rabinowitz is one of them, and a fashion shows up in a volume he edited, *Problems and Solutions from the Mathematical Visitor 1877–1896* (MathPro Press, Westford, Massachusetts, 1996). In *The Mathematical Visitor* at that time there was a vogue for problems of geometrical probability. An example of such a problem is Buffon's needle problem of calculating the probability that a needle dropped on a floor will intersect the line between floorboards. *The Mathematical Visitor* had many such, some of them elaborated to an extent that we would say is considerably too far. For example, here is a problem proposed by E. B. Seitz of Greenville, Ohio:

> A chord is drawn through two points taken at random in the surface of a circle. If a second chord be drawn through two other points taken at random in the surface, find the chance that it will intersect the first chord.

It is a fairly natural question, but the elaboration occurs in the answer, which is

$$\frac{16}{\pi^4 r^8} \int_0^{\pi/2} \int_0^\theta \int_0^{2\varphi} \int_0^{2\pi} \int_0^{2r\sin\theta} \int_0^w \int_0^{2r\sin\varphi} \int_0^y r\sin\theta\, d\theta\, r\sin\varphi\, d\varphi\, d\psi\, d\mu\, dw\, dx\, dy\, dz.$$

This simplifies to $\frac{1}{3} + \frac{245}{72\pi^2}$, or about .678, which seems reasonable.

Those kinds of problems have gone away, but they may be back some day. Some problems, especially those in recreational mathematics, go on and on. In the following paper, David Eugene Smith, a thorough scholar, traces the history of some very old problems. (I have omitted his numerous footnotes giving sources.)

⌘

On the Origin of Certain Typical Problems

David Eugene Smith (1917)

One thing which impresses the student of mathematical problems is that several which he would naturally classify as purely fictitious and of the nature of pleasing puzzles apparently had their origin in genuine applications of mathematics to questions of real life. Of these I shall mention only four, although the list could be greatly extended.

The first of these problems, without which an algebra of to-day might by some be thought to be incomplete, so rooted is it in our traditions, is that of the pipes filling the cistern. No problem has had a longer and more continuous history, and the traveler who is familiar with the Mediterranean lands cannot fail to recognize that here is its probable origin. Not a town of any size that bears the stamp of the Roman power is without its public fountain into which or from which several conduits lead. In the domain of physics, therefore, this would naturally be the most real of all the problems that came within the purview of every man, woman, or child of that civilization. Furthermore, the elementary clepsydra may also have suggested the same line of problems, the principle involved being the same.

The problem in definite form first appears in Heron's Μετρήσεις of about 100 AD, and although there is some question as to the authorship and date of the work, there is none as to the fact that this style of problem would appeal to such a writer as he. It next appears in the writings of Diophantus, c. 275 AD, and among the Greek epigrams attributed to Metrodorus, c. 325 AD, and soon after this it became common property in the east as well as the west. It is found in the list attributed to Alcuin (c. 825); in the great classic of India, the *Lilāvati* of Bhaskara (c. 1150); in the best-known of all the Arab works on arithmetic, the *Kholāsat-el-hisāb* of Behā-ed-din (1547–1622); and in numerous medieval manuscripts. When books began to be printed it was looked upon as one of the stock problems of the race, and many of the early writers gave it a prominent position, among them being men like Petzensteiner (1483), Tonstall (1522), Gemma Frisius (1540), and Robert Recorde (c. 1540).

Such, then, was the origin of what was once a cleverly stated problem of daily life. There is, however, this interesting law of book writers—that most of them will steal from one another without the least scruple if they can thinly veil the theft. This problem, therefore, like dozens of others, went through many metamorphoses, of which I shall mention only a few.

In the fifteenth century, and very likely much earlier, there appeared the variant of a lion, a dog, and a wolf, or other animals, eating a sheep, and this form was even more common in the sixteenth century.

In the sixteenth century we also find in various books the variant of the case of men building a wall or a house, in place of pipes filling a cistern, and this form has survived to the present time. It appeared in Tonstall's exhaustive treatise, *De Arte Supputandi*, in 1522 in Cataneo's well-known work of 1546, and in due time became modified to the form beginning, "If A can do a piece of work in 4 days, B in 3 days," and so on.

The influence of the wine-drinking countries shows itself in the variant given by that remarkable writer Gemma Frisius (1540), who states that a man can drink a cask of wine in 20 days, but if his wife drinks with him it will take only 14 days, from which it is required to find the time it would take his wife alone.

The influence of a rapidly growing commerce led one of the German writers of 1540 to consider the case of a ship with 3 sails, by the aid of the largest of which a voyage could be made in 2 weeks; with the next in size in 3 weeks, and with the smallest in 4 weeks, it being required to find the time if all three were used, several factors being evidently ignored, such as one sail blanketing the others and the speed not being proportional to the power

The agricultural interests changed it to a mill with four "Gewercken," and other interests continued to modify it further until, as is usually the case, the style of problem has tended to fall from its own absurdity. Merely mentioning one of our modern writers who modifies the problem to the case of the pipes of a gasoline tank in a motor car, I may close its varied history by referring to a writer of the early nineteenth century moved by a bigotry which we would not countenance in academic circles to-day, who proposed to substitute priests praying for souls in purgatory.

Thus we see a recreative problem, starting as an ingenuously worded practical case, becoming fictitious under changed conditions, maintaining itself for two thousand years because of its recreative feature, and almost falling by the wayside because of the absurdities which finally attached to it. It is likely to retain, however, some minor place in our schools because it is not only real within the imagination of pupils, which our technical mechanical problems usually are not, but it is interesting and illustrates a valuable mathematical principle.

The next problem to which I wish to call your attention has not maintained its place in our books although it has an honorable history of over 2,000 years;

it is interesting, it is real within the realm of the pupil's imagination; but it fails for the reason that no principle is involved that is needed in secondary mathematics. The problem is the one commonly known as the Josephsspiel, or the one of the Turks and Christians. It relates that 15 Turks and 15 Christians were on a ship and that half had to be sacrificed; it being necessary to choose the victims by lot, the question is as to how they can be arranged in a circle so that, in counting round, every fifteenth should be a Turk.

It is probable that the problem goes back to the custom of *Decimatio* in the old Roman armies, the selection by lot of every tenth man when a company had been guilty of cowardice, mutiny, or loss of standards in action. Both Livy and Dionysius speak of it in the case of the mutinous army of the consul Appius Claudius (BC 471), and Dionysius further speaks of it as a general custom. Polybius says that it was a usual punishment when troops had given way to panic. The custom seems to have died out for a time, for when Crassus resorted to decimation in the war of Spartacus he is described by Plutarch as having revived an ancient punishment. It was extensively used in the civil wars and was retained under the Empire, sometimes as *vicesimatio* (every twentieth man being taken), and sometimes as *centesimatio* (every hundredth man).

Now it is very improbable that those in charge of the selection would fail to have certain favorites, and hence it is natural that there may have grown up a scheme of selection that would save the latter from death. Such customs may depart, but their influence remains in various ways. In the present great war we have frequently read of a regiment being decimated; but how few of us have thought of the origin of the expression

In its semi-mathematical form it is first referred to in the work of an unknown author, possibly Ambrose of Milan, who wrote, under the nom de plume of Hegesippus, a work *De bello iudaico*. In this work he refers to the fact that Josephus, the author of the well-known history of the wars of the Jews, was saved on the occasion of a choice of this kind. Indeed, Josephus himself refers to the matter of his being saved by lucky chance or by the act of God

The oldest European trace of the problem, aside from that of Hegesippus, is found in Codex Einsidelensis No. 326, of the beginning of the tenth century. It is also referred to in a manuscript of the eleventh century now in the Munich library and in Codex Bernensis No. 704, of the twelfth century. It is given in the *Ta'hbula* of Rabbi ben Esra (d. 1167) in the twelfth century, and indeed it is to this writer that Elias Levita, who seems first to have given it in printed form (1518), attributes its authorship.

The problem, as it came to be stated, related that Josephus at the time of the sack of the city of Jotapata by Vespasian, hid himself with forty other Jews in a cellar. It becoming necessary to sacrifice some of the number, a method analogous to the old Roman method of *decimatio* was adopted, but in such way as to preserve himself and a special friend. It is on this account that the Germans still call the problem by the name of Josephsspiel.

Chuquet (1484) mentions the problem, as does at least one other writer of the fifteenth century. When, however, printed works on algebra and higher arithmetic began to appear, it became well known. The fact that such writers as Cardan and Ramus gave it prominence was enough to assure its coming to the attention of scholars.

Like so many curious problems, this one found its way to the Far East, appearing in the Japanese books as relating to a mother-in-law's selection of the children to be disinherited. With characteristic Japanese humor, however, the woman was described as making an error in her calculations so that her own children were disinherited and her step-children received the estate.

The third problem of which I think the origin is worth our attention is the common one of the testament. It relates that a man about to die made a will bequeathing 1/3 of his estate to his widow in case an expected child was a son, the son to have 2/3; and 2/3 to the widow if the child was a daughter, the daughter to have 1/3. The issue was twins, one a boy and the other a girl, and the question was as to the division of the estate.

The problem in itself is of no particular interest, being legal rather than mathematical; but I mention it because it is a type and is by no means isolated. Under both the Roman and the Oriental influence these inheritance problems played a very important role in such parts of analysis as the ancients had developed. In the year 40 BC the *lex Falcidia* required at least 1/4 of an estate to go to the legal heir. If more than 3/4 was otherwise disposed of, this had to be reduced by the rules of partnership. Problems involving this "Falcidian fourth" were therefore common under the Roman law, just as problems involving the widow's dower right were and are common in the English law and in this country.

The problem as I have stated it appears in the writings of Juventius Celsus, a celebrated jurist of about 75 AD, who wrote on testamentary law; in those of Salvianus Julianus, a jurist in the reigns of Hadrian (117–138), and Antoninus Pius (138–161), and in those of Caecilius Africanus (c. 100), celebrated for his knotty legal puzzles.

In the Middle Ages it was a favorite conundrum, and in the early printed arithmetics it is often found in a chapter on inheritances which reminds us of the Hindu mathematical collections. It went through the same later development that characterizes most problems and finally fell on account of its very absurdity. That is, Widman (1489) takes the case of triplets, one boy and two girls, and in this he is followed by Albert (1540) and Rudolff (1526). Cardan (1539) complicates it by supposing 4 parts to go to the son and 1 part to the mother, or 1 part to the daughter and 2 parts to the mother, and in some way decides on an 8, 7, 1 division. Texeda (1545) supposes 7 parts to go to the son and 5 to the mother, or 5 to the daughter and 6 to the mother, while other writers of the sixteenth century complicate the problem even more. The final complications of the "swanghere Huysvrouwe" or "donna grauida" are found in some of the Dutch books, and these and the change in ideas of propriety account for the banishment of the problem from books of our day. The most sensible remark about the problem to be found in any of the early books is given in the words of the "Scholer" in Robert Recorde's *Ground of Artes* (c. 1540): "If some cunning lawyers had this matter in scanning, they would determine this testament to be quite voyde, and so the man to die intestate, because the testament was made insufficient."

The fourth problem to whose origin and development I wish to direct special attention is the one of pursuit. It would be difficult to conceive of a problem that would seem more real, since we commonly overtake a friend in walking, or are in turn overtaken. It would therefore seem very certain that this problem is among the ancient ones in what was once looked upon as higher analysis. We have a striking proof that this must be the case in the famous paradox of Achilles and the Tortoise, the history of which has been so carefully and entertainingly worked out by our colleague, Professor Cajori. It is a curious fact, however, that it is not to be found in the Greek collections, although it must also be said that we have not a single work on the Greek *logistice* (λογιστικέ) extant, so that it may have been common without our knowing of the fact. It appears, however, among the *Propositiones ad acuendos juvenes* attributed to Alcuin, in the form of the hound pursuing the hare, and thereafter it was looked upon as one of the stock questions of European mathematics. I have run across it in an Italian manuscript of c. 1440, it is in Petzensteiner's work of 1483, Calandri used it in 1491, Pacioli gives it in his Suma of 1494, and most of the writers of any prominence in the sixteenth century embodied it in their lists.

In those centuries when commercial communication was wholly by means of couriers who traveled regularly from city to city, a custom still determining the name of *correo* for a postman in certain parts of the world, the problem of the hare and hound naturally took on the form of, or perhaps paralleled, the one of the couriers. This problem was not, however, always one of pursuit, since the couriers might be traveling either in the same direction or in opposite directions. This variant of the stock problem is purely Italian, for even the early German writers give it with reference to Italian towns. As a matter of course also, it was varied by substituting ships for couriers, while our modern textbook writers show their lack of originality by merely substituting automobiles for ships.

It was natural to expect that the problem should have a further variant, namely, the one in which the couriers should not start simultaneously. In this form it first appeared in print in Germany in 1483, in Italy in 1484, and in England in 1522.

The invention of clocks with minute hands as well as hour hands gave the next variant, as to when both hands would be together—a relatively modern form of the question, as is also the astronomical problem of the occurrence of the new moon. The latest form, however, has to do with the practical question of a railway timetable, but here graphic methods naturally take the place of analysis so that of all the variants those of the couriers and the clock hands seem to be the only ones that will survive. Neither is valuable *per se*, but each is interesting, each is real within the range of easy imagination, and each involves a valuable mathematical principle—a fairly refined idea of function, and so it is probable that each will persist in spite of the present transitory period of the attempted debasement of elementary mathematics.

David Eugene Smith (1860–1944) taught at Teacher's College, Columbia from 1901 until his death. He was a historian of mathematics and author of many books, including *Rara Arithmetica*, *History of Mathematics* (two volumes), translations, and even a book for children, *Number Stories of Long Ago*. His notes to the 1915 edition of De Morgan's *A Budget of Paradoxes* take up almost as much space as the original text, and are amazingly thorough and detailed. He was President of the Mathematical Association of America, 1920–21.

He taught at what is now Eastern Michigan University from 1891 to 1898 and earned his PhD degree at Syracuse. From 1898 to 1901 he was at Brockport (New York) State Normal School, first as a teacher and then as its Principal, as chief executive officers of teacher training schools evidently were called back then. He was of the opinion that teacher training in New York lagged behind the achievements of Michigan and asked the state legislature for more money for Brockport. He was turned down, so he quit. What would have happened if the state of New York had given him the $1500 (or however much it was) that he wanted? He might have stayed at Brockport. Though he no doubt would have retired, full of years and honors and, perhaps later, Brockport would name Smith Hall after him, it was for the best that he moved to the stimulation of New York City. Scholarship would have been the loser otherwise. Sometimes state legislatures can do the right thing, even if for the wrong reason.

14

A Tangled Tale

Charles Lutwidge Dodgson, or Lewis Carroll, (1832–1898) had many talents. He was first in mathematics in his Oxford class, he was a pioneering photographer, and he wrote *Alice's Adventures in Wonderland* (1865) and *Through the Looking-Glass* (1871). The details of his life and works are sufficiently well-known that they do not need to be repeated here.

He wrote a lot, works both serious and not, and even some of his lesser works are worth looking at. Charlotte M. Yonge, a novelist now forgotten but highly regarded in her day, was the editor of *The Monthly Packet* for its entire existence, 1851–1899. The journal, whose full title was *The Monthly Packet of Evening Readings for Younger Members of the English Church,* was aimed at girls and women of the middle and upper classes. *A Tangled Tale* appeared in the magazine in 1880 as a serial of ten "knots," each containing a problem or two, to which solutions were given in the next issue. Carroll said it was for "the amusement, and possible edification of the fair readers of that magazine." The tale had a negligible plot—the problems were the thing.

One commentator wrote

> *A Tangled Tale* is full of the linguistic fun for which Carroll is famous, but perhaps the most striking thing about it is the condescending tone he uses when pointing out the flaws in the reasoning of their arguments sent to him.

I don't see this myself—the *most* striking thing?—but I may have been teaching so long that I can no longer recognize condescension. Here is Knot II, with solution, so you may judge for yourself.

A Tangled Tale

Lewis Carroll (1880)

Knot II. Eligible Apartments

> "Straight down the crooked lane,
> And all round the square."

"Let's ask Balbus about it," said Hugh.

"All right," said Lambert.

"He can guess it," said Hugh.

"Rather," said Lambert.

No more words were needed: the two brothers understood each other perfectly.

Balbus was waiting for them at the hotel: the journey down had tired him, he said: so his two pupils had been the round of the place, in search of lodgings, without the old tutor who had been their inseparable companion from their childhood. They had named him after the hero of their Latin exercise-book, which overflowed with anecdotes of that versatile genius—anecdotes whose vagueness in detail was more than compensated by their sensational brilliance. "Balbus has overcome all his enemies" had been marked by their tutor, in the margin of the book, "Successful Bravery." In this way he had tried to extract a moral from every anecdote about Balbus—sometimes one of warning, as in "Balbus had borrowed a healthy dragon", against which he had written "Rashness in Speculation"—sometimes of encouragement, as in the words "Influence of Sympathy in United Action," which stood opposite to the anecdote "Balbus was assisting his mother-in-law to convince the dragon"—and sometimes it dwindled down to a single word, such as "Prudence," which was all he could extract from the touching record that "Balbus, having scorched the tail of the dragon, went away." His pupils liked the short morals best, as it left them more room for marginal illustrations, and in this instance they required all the space they could get to exhibit the rapidity of the hero's departure.

Their report of the state of things was discouraging. That most fashionable of watering. places, Little Mendip, was "chockfull" (as the boys expressed it) from end to end. But in one Square they had seen no less than four cards, in different houses, all announcing in flaming capitals "ELIGIBLE APART-MENTS." "So there's plenty of choice, after all, you see," said spokesman Hugh in conclusion.

"That doesn't follow from the data," said Balbus, as he rose from the easy chair, where he had been dozing over *The Little Mendip Gazette.* "They may be all single rooms. However, we may as well see them. I shall be glad to stretch my legs a bit."

An unprejudiced bystander might have objected that the operation was needless, and that this long, lank creature would have been all the better with even shorter legs: but no such thought occurred to his loving pupils. One on each side, they did their best to keep up with his gigantic strides, while Hugh repeated the sentence in their father's letter, just received from abroad, over which he and Lambert had been puzzling. "He says a friend of his, the Governor of—*what* was that name again, Lambert? " ("Kgovjni," said Lambert.) "Well, yes. The Governor of—what-you-may-call-it—wants to give a very small dinner-party, and he means to ask his father's brother-in-law, his brother's father-in-law, his father-in-law's brother, and his brother-in-law's father: and we're to guess how many guests there will be."

There was an anxious pause. "*How* large did he say the pudding was to be?" Balbus said at last. "Take its cubical contents, divide by the cubical contents of what each man can eat, and the quotient—"

"He didn't say anything about pudding," said Hugh,"—and here's the Square," as they turned a corner and came into sight of the "eligible apartments."

"It *is* a Square!" was Balbus' first cry of delight, as he gazed around him. "Beautiful! Beau-ti-ful! Equilateral! *And* rectangular!"

The boys looked round with less enthusiasm. "Number nine is the first with a card," said prosaic Lambert; but Balbus would not so soon awake from his dream of beauty.

"See, boys!" he cried. "Twenty doors on a side! What symmetry! Each side divided into twenty-one equal parts! It's delicious!"

"Shall I knock, or ring?" said Hugh, looking in some perplexity at a square brass plate which bore the simple inscription "RING ALSO."

"Both," said Balbus. "That's an Ellipsis, my boy. Did you never see an Ellipsis before?"

"I couldn't hardly read it," said Hugh, evasively. "It's no good having an Ellipsis, if they don't keep it clean."

"Which there is *one* room, gentlemen," said the smiling landlady. "And a sweet room too! As snug a little back-room— "

"We will see it," said Balbus gloomily, as they followed her in. "I knew how it would be! One room in each house! No view, I suppose?"

"Which indeed there *is*, gentlemen!" the landlady indignantly protested, as she drew up the blind, and indicated the back garden.

"Cabbages, I perceive," said Balbus. "Well, they're green, at any rate."

"Which the greens at the shops," their hostess explained, "are by no means dependable upon. Here you has them on the premises, *and* of the best."

"Does the window open?" was always Balbus' first question in testing a lodging: and "Does the chimney smoke?" his second. Satisfied on all points, he secured the refusal of the room, and they moved on to Number Twenty-five.

This landlady was grave and stern. "I've nobbut one room left," she told them: "and it gives on the back-garden."

"But there are cabbages?" Balbus suggested.

The landlady visibly relented. "There is, sir," she said: "and good ones, though I say it as shouldn't. We can't rely on the shops for greens. So we grows them ourselves."

"A singular advantage," said Balbus: and, after the usual questions, they went on to Fifty-two.

"And I'd gladly accommodate you all, if I could," was the greeting that met them. "We are but mortal," ("Irrelevant!" muttered Balbus) "and I've let all my rooms but one."

"Which one is a back-room, I perceive," said Balbus: "and looking out on— on cabbages, I presume?"

"Yes, indeed, sir!" said their hostess. "Whatever *other* folks may do, *we* grows our own. For the shops—"

"An excellent arrangement!" Balbus interrupted. "Then one can really depend on their being good. Does the window open?"

The usual questions were answered satisfactorily: but this time Hugh added one of his own invention —"Does the cat scratch?"

The landlady looked round suspiciously, as if to make sure the cat was not listening, "I will not deceive you, gentlemen," she said. "It *do* scratch, but not without you pulls its whiskers! It'll never do it," she repeated slowly, with a visible effort to recall the exact words of some written agreement between herself and the cat, "without you pulls its whiskers!"

"Much may be excused in a cat so treated," said Balbus, as they left the house and crossed to Number Seventy-three, leaving the landlady curtseying on the doorstep, and still murmuring to herself her parting words, as if they were a form of blessing, "—not without you pulls its whiskers!"

At Number Seventy-three they found only a small shy girl to show the house, who said "yes'm" in answer to all questions.

"The usual room," said Balbus, as they marched in: "the usual back-garden, the usual cabbages. I suppose you can't get them good at the shops?"

"Yes'm," said the girl.

"Well, you may tell your mistress we will take the room, and that her plan of growing her own cabbages is simply *admirable!*"

"Yes'm," said the girl, as she showed them out.

"One day-room and three bed-rooms," said Balbus, as they returned to the hotel. "We will take as our day-room the one that gives us the least walking to do to get to it."

"Must we walk from door to door, and count the steps?" said Lambert.

"No, no! Figure it out, my boys, figure it out!" Balbus gaily exclaimed, as he put pens, ink, and paper before his hapless pupils, and left the room.

"I say! It'll be a job!" said Hugh.

"Rather!" said Lambert.

Answers to Knot II

§1. The Dinner Party

Problem. "The Governor of Kgovjni wants to give a very small dinner party, and invites his father's brother-in-law, his brother's father-in-law, his father-in-law's brother, and his brother-in-law's father. Find the number of guests."

Answer. "One."

In this genealogy [on the right], males have been denoted by capitals, and females by small letters.

The Governor is E and his guest is C.

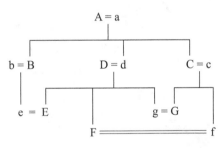

Ten answers have been received. Of these, one is wrong, GALANTHUS NIVALIS MAJOR, who insists on inviting *two* guests, one being the Governor's *wife's brother's father*. If she had taken his *sister's husband's father* instead, she would have found it possible to reduce the guests to *one*.

Of the nine who send right answers, SEA-BREEZE is the very faintest breath that ever bore the name! She simply states that the Governor's uncle might fulfill all the conditions "by intermarriages"! "Wind of the western sea," you have had a very narrow escape! Be thankful to appear in the Class-list at all! BOG-OAK and BRADSHAW OF THE FUTURE use genealogies which require 16 people instead of 14, by inviting the Governor's *father's sister's husband* instead of his *father's wife's brother*. I cannot think this so good a solution as one that requires only 14. CAIUS and VALENTINE deserve special mention as the only two who have supplied genealogies.

CLASS LIST

I.

BEE.   M. M.   OLD CAT

CAIUS   MATTHEW MATTICKS   VALENTINE

II.

BOG-OAK   BRADSHAW OF THE FUTURE

III.

Sea-Breeze

§2. *The Lodgings*

Problem. "A Square has 20 doors on each side, which contains 21 equal parts. They are numbered all round, beginning at one corner. From which of the four, Nos. 9, 25, 52, 73, is the sum of the distances, to the other three, least?"

Answer. "From No. 9."

Let A be No. 9, B No. 25, C No. 52, and D No. 73.

Then $AB = \sqrt{12^2 + 5^2} = \sqrt{169} = 13$;

$AC = 21$;

$AD = \sqrt{9^2 + 8^2} = \sqrt{145} = 12+$

(N. B., i.e., between 12 and 13.);

$BC = \sqrt{16^2 + 12^2} = \sqrt{400} = 20$;

$BD = \sqrt{3^2 + 21^2} = \sqrt{450} = 21+$;

$CD = \sqrt{9^2 + 13^2} = \sqrt{250} = 15+$.

Hence sum of distances from A is between 46 and 47; from B, between 54 and 55; from C, between 56 and 57; from D, between 48 and 51. (Why not "between 48 and 49"? Make this out for yourselves.) Hence the sum is least for A.

Twenty-five solutions have been received. Of these, 15 must be marked " 0," 5 are partly right, and 5 right. Of the 15, I may dismiss ALPHABETICAL PHANTOM, BOG-OAK, DINAH MITE, FIFEE, GALANTHUS NIVALIS MAJOR (I fear the cold spring has blighted our SNOWDROP), GUY, H. M. S. PINAFORE, JANET, and VALENTINE with the simple remark that they insist on the unfortunate lodgers *keeping to the pavement.* (I used the words "crossed to Number Seventy-three" for the special purpose of showing that short cuts were possible.) SEA-BREEZE does the same, and adds that "the result would be the same" even if they crossed the Square, but gives no proof of this. M. M. draws a diagram, and says that No. 9 is the house, "as the diagram shows." I cannot see how it does so. OLD CAT assumes that the house must be No. 9 or No. 73. She does not explain how she estimates the distances. BEE's Arithmetic is faulty: she makes $\sqrt{169} + \sqrt{442} + \sqrt{130} = 741$. (I suppose you mean $\sqrt{741}$ which would be a little nearer the truth. But roots cannot be added in this manner. Do you think $\sqrt{9} + \sqrt{16}$ is 25, or even $\sqrt{25}$?) But AYR's state is more perilous still: she draws illogical conclusions with a frightful calmness. After pointing out (rightly) that AC is less than BD she says, "therefore the nearest house to the other three must be A or C." And again, after pointing out (rightly) that B and D are both within the half-square containing A, she says "therefore" $AB + AD$ must be less than $BC + CD$. (There is no logical force in either "therefore." For the first, try Nos. 1, 21, 60, 70: this will make your premise true, and your conclusion false. Similarly, for the second, try Nos. 1, 30, 51, 71.)

Of the five partly-right solutions, RAGS AND TATTERS and MAD HATTER (who send one answer between them) make No. 25 6 units from the corner instead of 5. CHEAM, E. R. D. L., and MEGGY POTTS leave openings at the corners of the Square, which are not in the *data*: moreover CHEAM gives values for the distances without any hint that they are only *approximations*. CROPHI AND MOPHI make the bold and unfounded assumption that there were really 21 houses on each side, instead of 20 as stated by Balbus. "We may assume," they add, "that the doors of Nos. 21, 42, 63, 84, are invisible from the centre of the Square"! What is there, I wonder, that CROPHI AND MOPHI would *not* assume?

Of the five who are wholly right, I think BRADSHAW OF THE FUTURE, CAIUS, CLIFTON C., and MARTREB deserve special praise for their full *analytical* solutions. MATTHEW MATTICKS picks out No. 9, and proves it to be the right house in two ways, very neatly and ingeniously, but why he picks it out does not appear. It is an excellent *synthetical* proof, but lacks the analysis which the other four supply.

CLASS LIST

I.

BRADSHAW OF THE FUTURE	CLIFTON C.
CAIU	MARTREB

II.

MATTHEW MATTICKS

III.

CHEAM	MEGGY POTTS
CROPHI AND MOPHI	{ RAGS AND TATTERS
E. R. D. L.	{ MAD HATTER

"Balbus" means "stammerer"; Dodgson had a stammer, so the tutor may be he, making fun of himself.

The book appeared in a Dover edition in 1958 bound together with *Pillow Problems*, a collection of problems that Carroll said that he composed and solved mentally while in bed waiting for sleep. It is still in print, in several editions.

15
A Brief Life

I am fond of John Aubrey's *Brief Lives*, which is why the following selection appears here. Aubrey (1626–1697), as Edmund Wilson says in his introduction to a 1949 edition of the book,

> … loved to compile gossip about famous men and to note their peculiarities, and in pursuit of this information he often went to considerable trouble. It was said of him by one of his friends that he expected to hear of Aubrey's breaking his neck someday as the result of dashing downstairs to get a story from a departing guest.

Wilson also said

> You can see much of the seventeenth century through Aubrey, because he was involved more or less in almost all its activities and interests: social, political, religious, scientific, and literary. And Aubrey's own accounts of his contemporaries and of the men of the preceding period—Boyle, Harvey, Hobbes, Ben Jonson, Marvell, Milton, Monk, Raleigh, Shakespeare, Sidney, Rochester, Wolsey, etc.—takes you quite out of the Hall of Monuments of the established reputation and the accepted classic, and into a world where all these celebrities are still alive and kicking or have not long ceased to be so…. I have never read anything else that made me feel in quite the same way what it must have been like to live then, to find oneself part of an England that was venturesome, unsettled, and eager, that was opening new horizons.

Just so. That is one of the main purposes of history. Santayana's theorem that those who do not know history are doomed to repeat it is probably false, but its converse, that those who know history will not repeat it, is proved false every day. Thus history has few *practical* applications, but it can expand our horizons. It is all too easy to assume, unconsciously, that the world has always been, and will always be, essentially as it is now, but this is of course false, as history shows. Aubrey lets us get inside the England of the seventeenth century. Whether or not we would want to live there, it is a good place to visit.

Aubrey has notes about quite a few mathematicians, including Isaac Barrow, Henry Briggs, René Descartes, and William Oughtred. There follows what he had to say about John Pell, whose name is attached to the diophantine equation $x^2 - Ny^2 = 1$. The biographical information was included by the book's editor. I have modernized Aubrey's spelling.

Brief Lives

John Aubry (c. 1690)

John Pell

[Born 1611. Mathematician. M.A. of Trinity College, Cambridge, 1630. Professor of Mathematics at Amsterdam (1643) and at Breda 1646. Returned to England in 1652 and was employed by Cromwell as a diplomat in Switzerland from 1654 to 1658. Rector of Fobbing 1661–85. Vicar of Laindon 1663–85. D.D. Lambeth 1663. His mathematical reputation was great, but he accomplished little, and left nothing of moment. He died in poverty in 1685.]

John Pell, S. T. Dr., was the son of John, who was the son of John. His father died when his son John was but 5 years old and six weeks, and left him an excellent Library.

He went to School at the Free-school at Stenning, a Burrough town in Sussex, at the first Founding of the school. At 13 years and a quarter old he went as good a scholar to Cambridge, to Trinity College, as most Masters of Arts in the University (he understood Latin, Greek, and Hebrew) so that he played not much (one must imagine) with his schoolfellows, for, when they had play-days, or after school-time, he spent his time in the Library aforesaid.

Before he went first out of England, he understood these Languages (besides his mother-tongue) viz. Latin, Greek, Hebrew, Arabic, Italian, French, Spanish, High-Dutch, and Low-Dutch.

Anno Domini 1632 he married Ithamara Reginalds, second daughter to Mr. Henry Reginalds of London. He had by her 4 sons and 4 daughters borne in this order S., D., D., S., D., S., D., S.

In the year 1638 Mr. Theodore Haake came first to be acquainted with Mr. Pell by Mr. S. Hartlib's means, who having heard of his extraordinary parts in all kind of learning, especially the Mathematics, persuaded that the same might be far more usefully employed and improved for the public advancement of Learning, he never left soliciting and engaging friends here to persuade Mr. Pell instead of keeping School, as he then did in Sussex, to come up to London, where he soon got into great esteem among the most learned, both Natives and Foreigners, with whom he conversed. But he so minded and followed still the Cultivating of his more abstracting Studies, and naturally averse from suing or stooping much for what he was worthy of, it was a good while before he obtained any suitable place or settlement.

Mr. Haake recommended him once to my Lord Bishop of Lincoln (*quondam* Lord Keeper of England) who became very desirous to see the Man, inviting them of purpose to dine once with his Lordship for the freer discourse of all sorts of literature and Experiments, to get a touch and taste that satisfaction Mr. Pell could give him. Which proved so pertinent and abundant that my Lord put the question to him whether he would accept of a Benefice which he was ready, glad, and willing to bestow on him for his Encouragement. Mr. Pell thanked his Lordship, saying he was not capacitate for that, as being no Divine and having made the Mathematics his main study, for the great public need and usefulness thereof, which he had in a manner devoted himself to improve and advance to the utmost of his reach and abilities. Which answer pleased my Lord so well that he replied, Alas! what a sad case it is that in this great and opulent kingdom there is no public encouragement for the excelling in any Profession but that of the Law and Divinity. Were I in place as once I was, I would never give over praying and pressing his Majesty till a noble Stock and Fund might be raised for so fundamental, universally useful, and eminent Science as Mathematics. And thereupon his Lordship requested Mr. Pell to befriend him with his visits as often as he could spare time, promising him always a very hearty welcome. Yet Mr. Pell who was no Courtier came there no more.

In the mean time he communicated to his friends his excellent *Idea Matheseos* in half a sheet of paper, which got him a great deal of repute, both at home and abroad, but no other special advantage, till Mr. John Morian, a very learned and expert Gentleman, gave Mr. Haake notice that Hortensius, Mathematical Professor at Amsterdam, was deceased, wishing that their friend Mr. Pell might succeed. Sir William Boswell, his Majesty's ambassador in Holland, being here then, Theodore Haake conferred with him about it, who promised all his assistance; and between them, and by these two, a call was procured from Amsterdam for Mr. Pell, in 1643: and in May 1644 T. H. met him settled there on his return out of Denmark. Where he was, among others, dearly welcome to Gerardus Joannes Vossius.

And soon after his fame was much augmented by his refuting a large book of Longomontanus *Quadratura*, which caused the Prince of Orange (Henry Frederick) being about to erect an Academy at Breda, to borrow Mr. Pell from the Magistrate of Amsterdam, to grace his new Academy with a man of that fame for a few years. And there being comfortably stayed, the most learned of the then Parliament here, jealous that others should enjoy a countryman of their own, they never left offers and promises till they got him hither to be—they gave out—Professor Honorarius here. But the success proved soon deficient, and reduced him to much inconvenience, as having now a charge of a pretty large Family, viz. his wife with 4 or 5 children. And this continued till in 1654 Oliver, Lord Protector, sent him Envoy to the Protestant Cantons of Switzerland; resided chiefly at Zurich. He was sent out with the Title of Ablegatus, but afterwards he had order to continue there with the Title of Resident.

In 1658 he returned into England and so little before the death of Oliver Cromwell that he never saw him since he was Protector.

Memorandum that in his Negotiation he did no disservice to King Charles II, nor to the church, as may appear by his letters which are in the Secretary's Office.

Richard Cromwell, Protector, did not fully pay him for his business in Piedmont, whereby he was in some want; and so when King Charles II had been at home ten months, Dr. Sanderson, Bishop of Lincoln, persuaded him to take Holy Orders.

Gilbert Sheldon, Lord Bishop of London, gave Dr. Pell the scurvy Parsonage of Laindon cum Basseldon in the infamous and unhealthy (aguish) Hundreds of Essex (they call it Killpriest sarcastically) and King Charles the Second gave him the Parsonage of Fobing, 4 miles distant.

At Fobbing, seven curates died within the first ten years; in sixteen years, six of those that had been his Curates at Laindon are dead; besides those that went away from both places; and the death of his Wife, servants, and grandchildren.

Gilbert Sheldon being made Archbishop of Canterbury, John Pell was made one of his Cambridge Chaplains (He has 2 Oxford Chaplains and 2 Cambridge) and complaining one day to his Grace at Lambeth of the unhealthiness of his Benefice as abovesaid, said my Lord, I do not intend that you shall live there. No, said Pell, but your Grace does intend that I shall die there.

Lord Brereton was sent to Breda to receive the Instruction of this worthy Person, by his grandfather (George Goring, the Earl of Norwich) anno 1647, where he became a good proficient, especially in Algebra to which his Genius most inclined him and which he used to his dying day, which was 17 March, 1680: lies buried in St. Martin's church in the Fields.

I cannot mention this Noble Lord but with a great deal of Passion, for a more virtuous person (besides his great learning) I never knew. I have had the honor of his acquaintance since his coming from Breda into England. Never was there greater love between Master and Scholar then between Dr. Pell and this Scholar of his, whose death hath deprived this worthy Doctor of an ingenious Companion and a useful Friend.

Now by this time (1680) you doubt not but this great, learned man, famous both at home and abroad, has obtained some considerable Dignity in the Church. You ought not in modesty to guess at less then a Deanery. Why, truly, he is staked to this poor preferment still! For though the parishes are large, yet (Curates, etc., discharged) he clears not above 3-score pound per annum (hardly fourscore) and lives in an obscure lodging, three stories high, in Jermyn Street, next to the sign of the Ship, wanting not only books but his proper MSS. which are many.

He could not cringe and sneak for preferment though otherwise no man more humble nor more communicative. He was cast into King's Bench prison for Debt Sept 7, 1680.

In March 1682 he was very kindly invited by Daniel Whistler, M.D., to live with him at the Physicians College in London, where he was very kindly entertained; which the Dr. liked and accepted of, loving good cheer and good liquor, which the other did also; where eating and drinking too much, was the cause of shortening his days. About the middle of June he fell extreme sick of a Cold and removed to a grandchild of his married to one Mr. Hastings in St. Margaret's Churchyard, Westminster, near the tower, who now (1684) lives in Brownlow Street in Drury Lane, where he was like to have been burnt in his bed by a candle. Nov. 26, fell into convulsion fits which had almost killed him.

Dr. Pell has often said to me that when he solves a Question, he strains every nerve about him, and that now in his old age it brings him to a Looseness.

Dr. J. Pell was the first inventor of that excellent way or method of the marginal working in Algebra. He has said to me that he did believe that he solved some questions *non sine Divino auxilio* [not without divine aid].

Dr. Pell had a brother a Surgeon and Practitioner in Physic, who purchased an Estate of the Natives of New-York and when he died left it to his Nephew

John Pell, only son of the Doctor, who is a justice of the Peace in New York, and lives well. It is a great estate 8 miles broad and several miles long. Dr. Pell thought to have gone over to him.

Both his Parsonages are of the value of two hundred pounds per annum (or so accounted) but the Doctor was a most shiftless man as to worldly affaires, and his Tenants and Relations cousined him of the Profits and kept him so indigent that he wanted necessaries, even paper and Ink, and he had not 6d. in his purse when he died, and was buried by the Charity of Dr. Richard Busby and Dr. Sharp, Rector of St. Giles-in-the-fields and Dean of Norwich, who ordered his Body to lie in a Vault belonging to the Rector (the price is X pounds).

He died of a broken heart.

Poor Pell! He has continued to have misfortune even after death. Here is what E. T. Bell had to say about him, in *The Last Problem* (Simon and Schuster, 1961), another Brief Life:

John Pell is a classic instance of a man who was neither born great nor achieved greatness; he had such notoriety as accompanies mathematics thrust upon him after he was dead. Though his name occurs frequently in histories of mathematics, especially in the theory of numbers, Pell mathematically was a nonentity, and humanly an egregious fraud. It is long past time that his name be dropped from the textbooks. But there it is, and there it will stay until mathematical historians take account of some plain facts in the technical records of mathematics.

Pell came of a respectable family in Sussex, England, and had a passable education at a free school. Neither then nor in his mature life did he show the slightest sign of genius. By almost stupid diligence he earned his B.A. degree (1623) and his M.A. (1630); he was fairly competent in Greek and Latin. It is not clear how these modest attainments got him a professorship at the University of Amsterdam, where he lectured on, of all things, Diophantine analysis, about which he knew practically nothing. Next, the Prince of Orange, who knew even less than Pell about mathematics, prevailed upon the alleged mathematician to migrate to the new college at Breda. Pell's mathematical stock now rose rapidly: Oliver Cromwell, who knew even less mathematics than the prince, induced Pell to return to England, and insisted on endowing him with a handsome salary. It must be said that Pell himself was innocent of any flimflam in all this; he just kept his mouth shut and left his fortune to the ignorance and stupidity of his patrons. It is notoriously easy to make a mathematical sucker of a military man, a statesman or a politician, and that is what Pell made of the bullet-headed Oliver. One distinction let to another; Oliver sent his mathematical genius on a religious mission to Switzerland and tripled his pay. On Cromwell's death (1658), Pell, who had toadied to the royalists, consolidated his dubious position in the Restoration by taking holy orders and kissing the foot of Charles. He did not get the bishopric he coveted. Worthless or crackpot projects, such as a fantastic calendar reform, brought him further renown, if not honor. Either by dishonesty or incompetence he drifted into insolvency, and spent his last years sponging on the generosity of his tolerant and deluded friends—genius must be served. He was reported to have done great mathematics. Had he? His friends, who couldn't tell plus from minus, insisted that he had. Pell did not contradict them; he maintained a discreet silence and a mystifying inaccessibility. This is enough of his life.

My! What injury did Pell do Bell, to inspire such spleen? Bell says that Pell "toadied to the royalists"; Aubrey says "Mr. Pell who was no Courtier"; whom shall we believe? I think I will go with Aubrey, who was there at the time, instead of Bell. This is not the only place Professor Bell has made things up for the sake of a good story. He did this *all the time.*

This illustrates another of the other pleasures of history—weighing the evidence and coming to conclusions.

There is a new edition of *Brief Lives* on the market (David R. Godine, 1999) that may differ from my copy (University of Michigan Press, 1957). In any edition, reading it will make the reader a larger person.

16
Cardano

We all know Cardano's method for solving cubics. Or, to be more accurate, we all know that there *is* a Cardano's method for solving cubics. There is nothing to it, really. Take your cubic and eliminate the x^2 term, as from

$$x^3 + 3x^2 + 6x + 7 = 0$$

we get, on letting $x = y - 1$,

$$
\begin{aligned}
x^3 &= y^3 \;-\; 3y^2 \;+\; 3y \;-\; 1 \\
3x^2 &= 3y^2 \;-\; 6y \;+\; 3 \\
6x &= 6y \;-\; 6 \\
7 &= 7 \\
\hline
x^3 + 3x^2 + 6x + 7 &= y^3 + 3y + 3.
\end{aligned}
$$

Then make the clever substitution $y = z - \dfrac{1}{z}$:

$$
\begin{aligned}
y^3 &= z^3 \;-\; 3z \;+\; \frac{3}{z} \;-\; \frac{1}{z^3} \\
3y &= 3z \;-\; \frac{3}{z} \\
3 &= \phantom{z^3 - 3z + \frac{3}{z}\;} 3 \\
\hline
y^3 + 3y + 3 &= z^3 - \frac{1}{z^3} + 3.
\end{aligned}
$$

So, the equation we have to solve is

$$z^3 - \frac{1}{z^3} + 3 = 0 \quad \text{or} \quad z^6 + z^3 - 3 = 0,$$

which is a quadratic in z^3. We know how to solve quadratics, so we know z, and hence y, and hence x. A couple of substitutions conquers cubics.

Of course this is easy only after we have seen it done but thinking of it in the first place was another matter entirely. The complicated story of who had the idea in the first place and how Cardano's name came to be attached to it I will not go into. It is dealt with in many places: an electronic search for "Cardano and cubics" produced 299 references.

Solving the cubic was an event that has not been sufficiently noticed by historians. It is going too far to say that it *caused* the Renaissance, but it certainly helped it along. The ancient Greeks never solved cubics. The solution was something *new*. Knowing that it is possible to find new knowledge gives impetus to attempts to find it. When it was thought that the ancient Greeks had said the last word on every non-theological subject, there was no point in trying to go beyond them. After finding the solution of the cubic, the idea of progress could take firmer hold.

Cardano (1501–1576) was a fascinating person. He was a prolific writer who listed more than 200 books (one being *In Praise of Nero*) among his works. He was a physician, a philosopher, a gambler, a true renaissance man. He left among his works an autobiography. *De Vita Propria Liber* (translated into English as *The Book of My Life* by Jean Stoner and published in 1930) from which the following excerpts are taken. Cardano's time is far from ours, but I think his book lets us get as close to it, and to him, as it is possible to get. I don't think that I would have liked Cardano very much (nor do I think that he would have paid me any attention) and I know that I am glad not to have to have lived when and where he did. But it is good to know about such things.

It is striking that in his autobiography there is no mention of the cubic or, except for titles of books he wrote, any mention of mathematics at all. Cardano seems to have thought of himself as a doctor first and foremost. In any event, his mathematical achievements do not get a mention.

Much else does. The temptation to quote more and more from the book was hard to resist. Here is a portion.

The Book of My Life

Jerome Cardan (1575)

Chapter Five. Stature and Appearance

I am a man of medium height; my feet are short, wide near the toes, and rather too high at the heels, so that I can scarcely find well-fitting shoes; it is usually necessary to have them made to order. My chest is somewhat narrow and my arms slender. The thickly fashioned right hand has dangling fingers, so that chiromantists have declared me a rustic; it embarrasses them to know the truth. The line of life upon my palm is short, while the line called Saturn's is extended and deep. My left hand, on the contrary, is truly beautiful with long, tapering, well-formed fingers and shining nails.

A neck a little long and inclined to be thin, cleft chin, full pendulous lower lip, and eyes that are very small and apparently half-closed; unless I am gazing at something ... such are my features. Over the eyebrow of my left eye is a blotch or wart, like a small lentil, which can scarcely be noticed. The wide forehead is bald at the sides where it joins the temples. My hair and beard were blonde; I am wont to go rather closely clipped. The beard, like my chin, is divided, and the part of it beneath my chin always was thick and long, seeming to have a more abundant growth thereunder. Old age has wrought changes in this beard of mine, but not much in my hair.

A rather too shrill voice draws upon me the censure of those who pretend to be my friends, for my tone is harsh and high; yet when I am lecturing it cannot be heard at any distance. I am not inclined to speak in the least suavely, and I speak too often.

I have a fixed gaze as if in meditation. My complexion varies, turning from white to red. An oval face, not too well filled out, the head shaped off narrowly behind and delicately rounded, complete a picture so truly commonplace that several painters who have come from afar to make my portrait have found no feature by which they could so characterize me, that I might be distinguished. Upon the lower part of my throat is a swelling like a hard ball, not at all conspicuous, and coming to me as an inheritance from my mother.

Chapter Six. Concerning My Health

My bodily state was infirm in many respects: by nature; as the result of several cases of disease; and in the symptoms of weakness which displayed themselves.

My head is afflicted with congenital discharges coming at times from the stomach, at times from the chest, to such an extent that even when I consider myself in the best of health, I suffer with a cough and hoarseness. When this discharge is from the stomach, it is apt to bring on a dysentery and a distaste for food. More than once I believed I had had a touch of poison, but I shortly and unexpectedly recovered.

Another trouble was a catarrh or rheum of the teeth, through the effects of which I began to lose my teeth, several at a time, from the year 1563 on. Before that I had lost but one or two. Now I have fourteen good teeth and one which is rather weak; but it will last a long time, I think, for it still does its share.

Indigestion, moreover, and a stomach not any too strong were my lot. From my seventy-second year, whenever I had eaten something more than usual, or had drunk too much, or had eaten between meals, or eaten anything not especially wholesome, I began to feel ill. I have set forth a remedy for the foregoing in the second book of my treatise *On Guarding the Health*.

In my youth I was troubled with congenital palpitation of the heart, of which I was absolutely cured by medical skill. I had hemorrhoids, also, and the gout, from which I was so nearly freed that I was more frequently in the habit of trying to call it back when it was not present, than of getting rid of it when I had it.

I ignored a rupture, another weakness, in its early stages; but later, from my sixty-second year on, I greatly regretted that I had not taken care of it, especially since I knew it to be an inheritance from my father. In the case of this rupture, something worthy of note occurred: the hernia had started from either side, and although it was neglected on the left side, was eventually healed completely in that part by natural processes. The right side, more carefully treated with ligatures and other attentions, grew worse. A cutaneous itching annoyed me constantly, now in this part, now in that.

In 1536 I was overtaken with—it scarcely seems credible—an extraordinary discharge of urine; and although for nearly forty years I have been afflicted with this same trouble, giving from sixty to one hundred ounces in a single day, I live well. Neither do I lose weight—that I wear the same rings is evidence of this; nor do I thirst inordinately. Many others, seized that same year by a similar disease, and who did not seek a remedy, held out much longer than those who sought medical aid.

The tenth of these infirmities is an annual period of sleeplessness lasting about eight days. These spells come in the spring, in the summer, in the autumn, and in the winter; so that almost a whole month, rarely less, is spent yearly, and sometimes two. This I am wont to cure by abstaining from certain kinds of food, especially heavy food, but I do not diminish the quantity. This insomnia has never missed a year.

Several actual cases of sickness overtook me during my life. In the second month of my life I had the plague. The next serious illness occurred in or about my eighteenth year; I do not recall the exact date, other than that it happened in the month of August. I went almost three days without food, and spent the time wandering about the outskirts of the town, and through the gardens. When I returned home at nightfall, I pretended that I had dined at the home of my father's friend Agostino Lanizario. How much water I drank in those three whole days I cannot truthfully say. On the last day, because I was not able to sleep, my heart palpitated wildly and the fever raged. I seemed to be on the bed of Asclepiades in which I was incessantly swung upward and downward until I thought I should perish in the night. When at length I slept, a carbuncle, which covered the upper false rib of my right side, broke, and from it, at first, there was a scanty black discharge. Luckily, owing to a dose of my father's prescription which I swallowed four times a day, such a copious sweat broke out upon me that it drenched the bed and dripped down from the boards to the floor.

In my twenty-seventh year I was taken with the tertian fever. On the fourth day I was delirious, and on the seventh as well; on that day, also, I began to recover. Gout attacked me when I was at Pavia in my forty-fourth year, and when I was fifty-five I was troubled with daily fevers for forty days, at the crisis of which I was relieved of one hundred and twenty ounces of urine on October 13, 1555. In 1559, the year I returned to Pavia, I was taken with colic pains for two days.

The symptoms of weakness which attended my state of health were varied. To begin with, from my seventh year until I was almost twelve, I used to rise at night and cry out vaguely, and if my mother and my aunt, between whom I used to sleep, had not held me I often should have plunged out of bed. Likewise, my heart was wont to throb violently, but calmed down soon under the pressure of my band. To this was due the peculiarity of my breathing. Until I was eighteen, if I went out in the wind, particularly a cold wind, I was not able to breathe; but if I held my breath as soon as I became aware of the difficulty, normal respiration was quickly restored. During the same period, from the hour of retirement until midnight I was never warm from my knees down. This led my mother, and others as well, to say that I would not live very long. On some nights, however, when I had warmed up, I became entirely drenched with a sweat so abundant and hot that those who were told of it could scarcely believe it.

When I was twenty-seven, I took double tertian fever, which broke on the seventh day. Later, I had daily fever for forty days when I was fifty-four years old.

In November of my fifty-sixth year, from drinking a mild draught of squill wine, I was taken with dysuria, very acute in form. First I fasted thirty-four hours; later, twenty more. I took some drops of pine gum and cured myself.

It was my custom—and a habit which amazed many—when I had no other excuse for a malady, to seek one, as I have said, from my gout. And for this reason I frequently put myself in the way of conditions likely to induce a certain distress—excepting only that I shunned insomnia as much as I could—because I considered that pleasure consisted in relief following severe pain. If, therefore, I brought on pain, it could easily be allayed. I have discovered, by experience, that I cannot be long without bodily pain, for if once that circumstance arises, a certain mental anguish overcomes me, so grievous that nothing could be more distressing. Bodily pain, or the cause of bodily distress—in which there is no disgrace—is but a minor evil. Accordingly I have hit upon a plan of biting my lips, of twisting my fingers, of pinching the skin of the tender muscles of my left arm until the tears come. Under the protection of this self-chastisement I live without disgracing myself.

I am by nature afraid of high places, even though they are extensive; also, of places where there is any report of mad dogs having been seen.

At times I have been tormented by a tragic passion so heroic that I planned to commit suicide. I suspect that this has happened to others also, although they do not refer to it in their books.

Finally, in my boyhood, I was afflicted for about two years with indications of cancer. There appeared, by chance, a start upon the left nipple. The swelling was red, dark, hard, and eating. Some swollen veins seemed to remove this toward my young manhood, and in that period a palpitation of the heart—before mentioned—succeeded the varices. From this cancerous growth came blood-blisters, full of blood, and an itching and foulness of the skin; and subsequently I was healed, contrary to all hope of any relief, by a natural sloughing of the mass of diseased skin, although I had removed some of the affections by means of medication.

Chapter Nineteen. Gambling and Dicing

Peradventure in no respect can I be deemed worthy of praise; for so surely as I was inordinately addicted to the chess-board and the dicing table, I know that I must rather be considered deserving of the severest censure. I gambled at both for many years, at chess more than forty years, at dice about twenty-five; and not only every year, but—I say it with shame—every day, and with the loss at once of thought, of substance, and of time.

Nor was the smallest ground for defense left me. If, nevertheless, anyone may wish to rise in my defense, let him not say that I had any love for gambling, but rather that I loathed the necessities which goaded me to gambling —calumnies, injustices, poverty, the contemptuous behavior of certain men, the lack of organization in my affairs, the realization that I was despised by many, my own morbid nature, and finally the graceless idleness which sprang from all these. It is a proof of the foregoing assertion that once I was privileged to act a respectable part in life, I abandoned those low diversions. Accordingly it was not a love of gambling, not a taste for riotous living which lured me, but the odium of my estate and a desire to escape, which compelled me.

Although I have expounded many remarkable facts in a book on the combinations in chess, certain of these combinations escaped me because I was busy with other occupations. Eight or ten plays which I was never able to recapture, seemed to outwit human ingenuity, and appeared to be stalemates.

I have added these remarks to advise any who may by chance happen upon the same extraordinary situations—and I hope somebody may—so that they may add their jot or tittle to the solution.

Chapter Twenty-nine. Journeys

In the course of my many journeys I have visited almost the whole of Italy, excepting Naples, Apulia, and the regions neighboring these. Likewise I have seen Germany, especially lower Germany, Switzerland and Tyrol; besides these countries, I have been in France, England, and Scotland; and I will tell how this happened.

John Hamilton, Archbishop of St. Andrews, was a chief of state in Scotland. He was natural brother to the regent, Pontifical legate and primate. He suffered with periodic attacks of asthma, recurring, in the beginning, at somewhat lengthy intervals. When he had reached his fortieth year, the interval was reduced to eight days, and death seemed imminent, inasmuch as in twenty-four hours, he was eased by no, or at best by very slight, relief. He repeatedly consulted, but in vain, the court physicians of Charles V, then Emperor, and of Henry, King of France. Finally, my name having come to his ears, he sent, through the intermediation of his own physician, two hundred crowns to me in Milan, to the end that I should proceed to Lyons, or to Paris at most, implying that he would come there. I, since I was not at that time engaged in lecturing, as I have explained above, willingly welcomed his terms.

And so in the year 1552 on February 23, I was ready to set out upon my journey, crossing by way of Domo d'Ossola, Sion, Geneva, and Mount Simplon;

and, leaving Lake Geneva behind, I arrived in Lyons on the 13th of March. It was during the Milanese carnival, on the sixth day, by common reckoning.

There I tarried forty-six days, but not a glimpse of the Archbishop, nor the physician himself, whom I was awaiting. Meanwhile, my fees for services much more than paid my expenses. Lodovico Birague, the distinguished Milanese citizen, with whom I had long maintained an intimate friendship, was there, serving at that time as captain of the royal infantry. He went so far as to come to me with an offer of a yearly stipend of a thousand crowns if I would enter the service of Marshal Brissac. During this time, William Casanate, physician to the Archbishop, arrived, bringing with him an additional three hundred crowns which he gave me in order that I might be encouraged to proceed to Scotland. He offered to defray all my traveling expenses thither, and made promises of additional liberal rewards.

Accordingly I was conducted across the country, by way of the River Loire, to Paris. There I happened to see the great Orontius, but he refused to visit me.

Nicholas LeGrand took me to see the private treasure vaults of the King of France at the church of St. Denis, a place of no very great fame, but rather the more noteworthy in my estimation especially because of the perfect horn of a unicorn preserved therein.

Thereafter, a conference was held with the physicians to the King. We dined together, but they did not succeed in getting an expression of my views at table, because before the meal they had wished me to take precedence in expressing my opinion.

I continued my journey on the best of terms with Jean Fernel, Jacques de la Boë and another court physician, all of whom I left regretfully. I went on to Boulogne in France; whence, escorted by fourteen armed riders, and twenty soldiers, at the order of the Prince of Sarepont, I traveled on to Calais. There I saw the tower of Caesar still standing.

Therefrom, having crossed an arm of the sea, I went to London. At length I came to Edinburgh, to the side of the Archbishop, on the 29th of June. I remained there until the 13th of September. For my services I received another 400 gold crowns, one neck-chain worth 25 crowns, a thoroughbred riding-horse, and many other gifts, so that not a single member of my party went away empty-handed.

Returning, I came first to Brabant, and in the region about Tongres I visited Gravelines, Antwerp, Bruges, Ghent, Brussels, Louvain, Malines, Liege, Aix-la-Chapelle, Cologne, Coblenz, Kleve, Andernach, Mainz, Worms, Spires, Strasburg, Basle, Neustadt, Berne, Besancon, crossing through the heart of the

Tyrolese state, and visiting Chur and Chiavenna, cities of this region. Finally, crossing Lake Como, I reached Milan on January 3, 1553.

Of all these places, I tarried only in Antwerp, Basle, and Besancon. My friends in Antwerp made every effort to retain me.

At London I was granted an audience with the King, and while there accepted one gift of an hundred crowns, and rejected another of five hundred—some say a thousand, the truth of which I am not able to ascertain—because I was not willing to give acknowledgment to a title of the king in prejudice to the Pope. While in Scotland, I became an intimate friend of the Duc du Cell, Viceroy of the French.

At Basle I almost took lodging in a hostelry infested with the plague; and, indeed, would have done so, had I not been warned by Guglielmo Gratarolo. At Besançon I was hospitably received by the Prelate of Lisieux, as I have already noted; he presented me with gifts, as was the case elsewhere.

Altogether, I have lived four years at Rome. I spent nine at Bologna, three at Padua, twelve at Pavia, four at Moirago—the first four of my life—and one at Gallarate. I was in the town of Sacco almost six years, and in Milan thirty-two. I made three successive moves within three years.

Besides my long journey to Scotland, I went to Venice and Genoa, and visited the cities which are on the way—Bergamo, Crema, Brescia and others. I went, as well, to Ferrara and to Florence, and beyond Voghera and Tortona.

Briefly, I may say I am familiar with nearly the whole of Italy excepting the Kingdom of Naples and states adjoining, as the old Apulian region, ancient Latium, the Marches, Umbria, Calabria, ancient Magna Grecia and Lucania, and Abruzzi.

But, you will ask, to what purpose is this account of all these cities? There is great value: for if you will look about you only a single day, according to the suggestion of Hippocrates, you will know what may be the nature of the place and the customs of the inhabitants, what section of the city it is better to choose, and what diseases are prevalent. We may also determine which of the regions we visit is more favorable, for at one time an entire district will prove scarcely habitable because of the cold, and again another district undesirable on account of troublesome times.

An acquaintance with other lands is, besides, profitable for the better understanding of history, and especially helpful to mathematicians writing treatises on geography, or to anyone interested in the nature and productive usefulness of plants and animals. Again it gives one a knowledge of the routes mainly traveled; and out of the experiences of travel many books on this subject are published in Italian, thus presenting additional information about things far distant.

Chapter Thirty. Perils, Accidents, and Manifold, Diverse, and Persistent Treacheries

The accidents which I am about to relate happened to me when I was residing in the house of the Catanei at Pavia. One morning I was going to the University halls. Snow lay upon the ground. I had paused to relieve myself beside a ruined wall on the right of the school premises; and therefrom, as I continued my way through the lower part of the passage, a loosened tile tumbled perpendicularly. I avoided this peril only because I could not walk through the snow which filled the upper pathway, next the wall, where my companion had urged me to go.

In the following year, 1540, if I am not mistaken, when I was passing through the Via Oriental, it suddenly occurred to me, for no reason whatsoever, to cross from the left side to the right; and when I was once over, a great mass of cement fell from a very high cornice on the opposite side, precisely such a distance ahead of me that certainly, unless I had changed my course, it would have ground me to bits; thanks to God, I escaped.

Shortly after, near the same place, I, riding a mule, passed near a large wagon. I wanted to proceed toward the right, for my business was urgent, and the delay was annoying. I said to myself, "What if this wain overturns!" And before long, even as I was stopping there, it did fall, and there is no doubt that it almost crushed me. I suffered visible harm, and had run no small risk.

It is not on account of this single incident that I marvel at the outcome; but because on so many occasions I have changed the direction of my going, always involuntarily, except in perils of this sort, or perhaps in other dangers I have not been aware of. Nevertheless, it is not the significant event which ought to be wondered at, but rather the frequent recurrence of similar instances.

When I was a boy of eleven, if I am not mistaken, I was entering the courtyard of Donato Carchani, a citizen of rank, when a small furry dog bit me in the abdomen. I was injured in five places, not seriously; but the wounds were blackish spots, and in this accident I can say nothing less than that I was exposed to I know not what dangers of rabies. If this had happened to an older person, what the bite had not done, fright would surely have finished completely.

In 1525, the year in which I became rector, I was almost drowned in Lake Garda. Rather reluctantly I had boarded a craft which was transporting some horses for hiring. During the crossing, the main mast, the rudder, and one of the two oars with which the boat was manned, were broken. The sail was rent, and even a smaller mast snapped off; and at length night overtook us. I reached

Sirmione in safety after the other passengers had abandoned even their faintest hopes, and I was all but desperate.

Had our embarkation been delayed but the fortieth part of an hour, we should have perished; for such a violent storm gathered that the iron bars of the window-shutters on the inn were bent. Though I had seemed thoroughly uneasy from the first, I contentedly ate a hearty supper once an excellent pike appeared on the table. The rest of the passengers were not so calm, excepting the counselor of our disastrous voyage, who had, however, done efficient service in the moment of our peril.

Once when I was in Venice on the birthday of the Blessed Virgin, I lost some money while gambling; on the following day I lost the rest, for I was in the house of a professional cheat. When I observed that the cards were marked, I impetuously slashed his face with my poniard, though not deeply. There were in the room two youths, the body servants of my adversary; two lances were fastened to the beamed ceiling; the key was turned in the door.

When, however, I had begun to win and had recovered all the money, his, as well as my own, and the clothes and rings which I had lost on a previous day, but from the beginning of the next day had won back from the start; and since I had earlier despatched these belongings to my lodging with my servant, I tossed a part of the money back, willing to make amends when I saw I had wounded him. Then I attacked the house-servants, but since they were unable to handle weapons, and were beseeching me to spare their lives, I let them off on the condition that they should throw open the door of the house.

The master, seeing such a commotion and tumult in his household, and anxiously fearing every moment's delay, I judge, because he had defrauded me in his own house with his marked cards, after making a rapid calculation of the slight difference between what he had to gain or what to lose, ordered the door to be opened; thus I escaped.

On that same day about eight o'clock in the evening, while I was doing my best to escape from the clutches of the police because I had offered violence to a Senator, and keeping meanwhile my weapons beneath my cloak, I suddenly slipped, deceived in the dark, and fell into a canal. I kept my presence of mind even as I plunged, threw out my right arm, and, grasping the gunwale of a passing boat, was rescued by the passengers. When I scrambled aboard the skiff, I discovered in it, to my surprise, the Senator with whom I had just gambled. He had the wounds on his face bound up with a dressing; yet he willingly enough brought me out a suit of garments such as sailors wear. Dressed in these clothes, I traveled with him as far as Padua.

Once in Antwerp when I went to a shop to purchase some gems, I stumbled into a pit, which was in the shop for some reason or purpose I did not understand, and hurt myself. My left ear was scratched, but I made light of the accident, for I suffered but an abrasion of the skin.

In the year 1566 while in Bologna, I was thrown from a careening carriage which could not be checked. In this accident the ring-finger of my right hand was broken, and my arm so badly injured that I was not able to bend it backwards. The stiffness remained for some days, and then passed over to the left arm, while the right remained unimpaired—a rather remarkable circumstance. But what is still more surprising is that nine years later, this stiffness, for no manifest reason, but rather like an omen, returned again to the right side where it troubles me at present. The finger, however, although I gave it no special treatment, had so healed that I suffer no inconvenience from it and am no longer annoyed by the crook in it.

What should I say about the risk I ran of plague in 1541? I had been summoned to attend the servant of a colonel, a Genovese gentleman of rank from the Island. The servant was in the grip of the plague, for he had just come from Switzerland where he had slept between two men, both plague-stricken, who had later died of it. Unaware of such serious circumstances, I bore, in my official capacity as Rector of the College of Physicians, the canopy of the Emperor on the occasion of his entry into Milan. When it was known that my patient at this very time had the plague, the colonel wanted to hide the evidence by hurriedly concealing the dead man—for he was considered as good as dead—in a place without the city. I was unwilling to give my assent to this concealment, fearing nothing more than to be the victim of a deception. But with the help of God the sick man recovered, contrary to all hope, and aided not a little by my almost fatherly care.

I scarcely know how to relate the most amazing occurrence which befell me in the year 1546. I was coming away, all care-free, from a house in which, on the previous day, a dog had snapped at me. He had not, however, actually bitten me to wound; but because he had sneaked upon me so slyly and quietly, I feared that he might have been a mad dog. Although he did not drink from a dish of water I set down for him, still, he did not flee from it, and he had eaten a chicken thigh which I had directed to be given him.

Well then, to continue: as I was coming along, I saw a very large dog approaching me, but still at a distance. It was on the Feast of Santa Cruz, in the month of April, and I was riding through an exceedingly pleasant way, closed in on either side by greening hedges and trees. I said to myself, "What affair have I with dogs? A dog yesterday; a dog today! I have never entertained silly fears, but who knows whether this dog is really mad?"

While I was considering these things the dog was loping along, straight for my mule's head, so that I hardly knew what to do next; as soon as he was close enough, he sprang forward, ready to attack me. I was mounted on a small mule, but immediately the idea of making the only move that could help matters occurred to me: I bent my head low upon the mule's neck, and the dog passed over my head, gnashing his teeth! He not only did not wound me, but did not even touch me—a fact which surely can be numbered among the miracles!

If I had not committed this story to writing by frequent reference to it in many of my books, I should think I had dreamed it. Or I should even suppose that I had suffered an hallucination brought on by my meditation, had I not by chance observed, in looking behind me, that the dog was running back and leaping at a boy who was following me on the left near the hedge. I returned and questioned the boy. "Tell me, I beg you"—for the dog had now gone on in headlong course—"whether you saw what that dog did? And has he injured you?"

"He did not hurt me in the least," replied the boy, "but I certainly saw what the brute did to you."

"Tell me, I beg you, what he did," I requested again.

"He sprang directly toward your head, and when you bent your head, cleared you without hurting you."

Then I said to myself, "For a truth, I am not wandering in my mind."

Yet this incident still seems beyond credence to all.

In general, then, I had four extremely perilous encounters. In a word, such hazards, that if I had not in some way circumvented them, it would have been a question of my life. The first was the danger of drowning; the second, mad dogs; the third and lesser because it was only an incipient danger, the falling mass of masonry; and last, the quarrel in the house of the Venetian noble.

I was the victim of just as many great discouragements and obstacles in my life. The first was my marriage; the second, the bitter death of my son; the third, imprisonment; the fourth, the base character of my youngest son. Thus systematically ought one's life to be examined.

I pass over the sterility of my daughter; the long strife with the College of Physicians; so many iniquitous, not to say actually unjust, implications brought against me; such vile uncertainty of bodily health and constant weakness; the fact that I had no associate whose friendship brought me good counsel and advantage. Had such a companionship not been lacking in my life, I might have been relieved in a large measure, and been saved the sting of many difficulties.

Attend now, while I relate certain episodes which, *mirabile dictu*, are concerned with a deadly plot; for it is, in truth, a curious story.

In the course of my teaching at Pavia, I was used to reading at home. At that time I had in the house a woman who rendered irregular service, the youth

Ercole Visconte, two young boys, and a man servant, I think. One of the boys was an amanuensis and a musician; the other, a foot-page.

The year 1562 brought me to the place where I had determined to leave Pavia, and resign my professorship. The Senate viewed this decision with manifest disfavor, as if the case involved a man acting in anger. Now it happened that there were two certain doctors in Pavia. One was my former pupil, a crafty man; the other, a professor extraordinary of medicine. He was a guileless man, and not such a one as I deem mischievous; yet, forsooth, what does greed for position and power not accomplish, especially when it is allied to a very genuine zeal for achievement? There was also, moreover, that illegitimate opponent. These rivals were striving their utmost to force me to quit the city, ready to adopt any measures for the sake of fulfilling their avowed purpose. Accordingly, when they despaired of getting rid of me because of the Senate's objection, as against my own eagerness to withdraw, these adversaries plotted to murder me. This they arranged to do, not with a sword, for fear of scandal and the Senate, but by a deadly scheme. My competitor saw that it was impossible for him to be head unless I should withdraw. Therefore they went about their business indirectly.

First they composed a letter in the name of my son-in-law—a most infamous and vile letter to which they did not neglect to add, as well, my own daughter's name—declaring that he was ashamed of acknowledging our relationship; that, in behalf of the Senate and the College he was ashamed of me; and that he regarded the situation such that my colleagues ought to feel it a disgrace for me to continue to be a teacher, and should take action for my removal.

I was aghast at the impudence of this bold attack from my own children; I had no idea what I should do, or say, or reply. How could I interpret these statements?

Now, however, it is clear that this so shameless and uncivil act had sprung from that nest of intrigue whence other similar plots were hatched; for within a few days a second letter appeared, signed by Fioravanti. This was the tenor of it: *That he was ashamed of me for the sake of his country, the college, and the faculty; that the rumor was being circulated everywhere that I was using my boys for immoral purposes; and that not satisfied with one, I had added another to my household—a state of affairs absolutely unprecedented; he asked that in the name of many interested persons I should have regard for the notorious scandal I was causing, since nothing else was discussed in Pavia save these plans for ousting me; that houses of certain citizens where these infamies were being committed were to be publicly designated.*

When I had read these words I was dumbfounded, unable to believe that this letter was his work—the work of a friend and a sober man. Yet the memory of that former letter from my son-in-law—as I supposed—still rankled. And now I gave up all idea that it was ever his composition, that he had ever devised such a thing; since, from that time to this day, whatever his mood—amiable or vexed—he has been steadfastly devoted to me, nor ever shown a trace of any adverse attitude toward me, to say nothing of an absurdity like that letter.

I pass over further consideration of why a man in other respects prudent should, even had he believed such tales, have committed them to a letter which might fall into the hands of any number of people; how he could have charged his father-in-law with a crime, uncertain as he must have been of it at best, and a crime so filthy, so base and so certain to result seriously for him.

I called for my cloak and went to Fioravanti. I questioned him about the letter; he confessed that he had written it! Then was I even more astounded, for I had absolutely no suspicions of treachery, much less any reason for harboring such a thought. I began to urge him for his motives and to inquire where under heaven the so well-known designs for ejecting me were being concocted. At this he began to hesitate, and knew not what to respond. He could talk only vaguely about notoriety and the Rector of the Gymnasium; the latter was entirely in favor of Delfino.

Nevertheless, when the latter perceived that the situation began to look as if he himself rather might be drawn into serious difficulty than I ensnared in a suspicion of crime, he changed his plans, and withdrew; for he well realized, for all his simplicity, in what a grave situation he was involved. Accordingly, from that day all these intrigues died away, and all their well-devised inventions tottered and fell.

I may explain, in addition, since I discovered it afterwards, how the affair was concocted. The Wolf and the Fox had persuaded the Sheep that if I had not been there the Senate by all means would have appointed him to a professorship at Pavia; he stood second in line. My competitor was thereupon arrogating to himself, under a certain ancient and customary tradition of succession, the right of retaining the first place as his own—Fox that he was! And it did not turn out that way as later events were witness.

The curtain having thus descended on the first act of the tragedy, the second was begun, in which some of the mysteries of the first were cleared. Above all, my enemies took care that the one who was destined to be such a disgrace to his country, his family, the Senate, the medical associations of Milan and Pavia, the faculty of professors, and finally to his pupils, should be invited to enter the Accademia degli Affidati at Pavia of which many learned theologians, two princes, the Duke of Mantua, and Marquis Pescara were already members. When my enemies observed that it did not please me to be thus drawn in, they endeavored to inspire me with apprehension. What was I to do? Disheartened by the death of my son and tried by every adversity, I at length acquiesced, particularly since the Senate would free me within a period, already definitely determined, from my duty of teaching in the Pavian Gymnasium. Not even then was I aware of their trick, nor thought it strange that they should desire association with him whom all the powers not fifteen days earlier had wished to proscribe—the monstrous spouse of all young boys ... O faith of God! O savage hearts of mortals! O bitter hatred of infamous and false friends! O shameless cruelty, more fell than many serpents!

What more did they devise? In the very entrance of the halls of the Accademia degli Affidati, I noticed a beam so placed that it would be easy for it to fall and kill anyone entering heedlessly. Whether this was done by chance or by purpose I know not; but it is certain that I invented excuses for appearing as seldom as possible at the Accademia, unless unseasonably or unexpectedly, like a mouse on the look-out for a trap. But nothing came of it, whether because they deemed it unwise to perpetrate a foul deed so publicly, or because they had intended nothing at all by it, or because they had abandoned it for devising other schemes.

For one thing, when, within a few days, I was called to attend the sick son of Pietro Marco Trono, a surgeon, they had hoisted above the doorway a lead weight which had been arranged to give the appearance of a device for holding the reed-curtains back. However it may have been, or how, or by what art it was so suspended that it would fall, I never investigated. But it fell, and had it touched me, all would have been over with me. By what a narrow margin I escaped only God knows. From this time on I began to be filled with vague apprehensions, not knowing exactly what I had to fear, so greatly was I confounded.

Now give ear to the third act wherein occurs the denouement. A short time had elapsed when that Sheep—Fioravanti—came to me asking whether I would be willing to permit the two boy musicians who lived with me to take a part in the celebration attending the singing of a new Mass. Now my tormentors, knowing that these boys were my cup-bearers and pregustators, had made arrangements with my maid-servant to give me poison. Shortly before this, in truth, they had asked Ercole whether he would be interested in taking a part in the celebration; and he, suspecting no evil, had promised; but when he saw that the boys also were being invited, he began to feel uneasy. Therefore he replied that only one of the boys was a musician, not two. Fioravanti, accordingly, being a clumsy man and carried away with his zeal for getting the boys out of the house, said, "Give me both of them, for we know that he is a musician, and although he may be untrained, he will help with the others to fill the chorus of boys." Ercole replied to the two who had come, "Permit me, sirs, to talk it over with the master."

Then he came in to me and exposed their designs, with an insight so clever, that, if I had not been a lunatic or a dolt, I should easily have divined what the villains were about, although until then I had not noticed. Upon Ercole Visconte's advice, and after he had insisted that the one boy did not know even a note of music, I resolved not to let the boys go,

Not two weeks later—perhaps a little longer—these same men had returned, and inquired whether I would like to give those boys permission to take a part in a comedy. It was at this time that Ercole hurried to me, saying, "Now the story is out; they wish to remove all your servants from your table in order to destroy you with poison. Not only must you keep alert and on the look-out for treachery of this sort, but you must look into every occasion, for there is no doubt that they are seeking earnestly for your destruction."

I said, "I think as much." And yet I was not able to bring my mind to accept such a notion. "What reply shall I make?" I continued to Ercole. "Simply tell them that you cannot spare your servants." I did so, and they went away.

At length, I made guess that, after much deliberation, it was decided to do away with me entirely.

On a Saturday—June 6th, if I am not mistaken—I was suddenly awakened from sleep near midnight. I found that a ring in which there was a setting of sapphire, was missing from my finger. I called the servant to rise and search, but he sought in vain. I myself rose and ordered him to fetch a light. He went to obey, but returned saying that there was no fire. I chided him severely, and ordered him to look again. Whereupon he came back, joyfully bearing fire in the tongs, or at least an ember which had remained red hot because of its size. Since he said there was no other, I said it would do; I ordered him to blow on it. He blew three times, but seeing no hope of getting flame, was taking the candle away from the coal when a tongue of flame leaped out and lighted the candle.

"Did you notice that, Giacomo Antonio?" I cried, calling the lad by his full name.

"Indeed I did," he responded.

"What exactly happened?" I urged.

"The candle took light, although the coal gave off no flame!"

"Take care then," said I, "lest it be extinguished. Let us seek the ring."

It was found on the floor directly under the middle of the bed, a point which it could scarcely have reached unless it had been flung against the wall with great force, and bounded back.

I vowed that I would not set foot out of the house on the following day. Occasion favored my vow, for it was a holy day and I had no patient. But in the morning four or five of my pupils, accompanied by Zaffiro, appeared and invited me to be present at a supper at which all the professors of the Gymnasium and members of high rank in the Accademia were to be guests. I explained that I was unable; but they thought I was unwilling to make part of the company because, as they knew, I did not dine formally. So they said, "For your sake we have postponed the banquet to the supper hour."

Again I asserted, "I can by no means accept."

They asked my reason, and I told them of the portent and the vow, so that all were astonished. But two of them kept exchanging glances, and asked several times whether I would not be loath to mar the pleasure of the banquet by my absence. I replied that I would rather do that than break my vow. An hour later, they returned and pressed me somewhat more urgently; but I persisted in my refusal; for had I not decided not to leave the house at all? However, during the evening, although the sky was overcast, I went to attend a poor victualler who was sick, since my vow did not forbid a professional call.

Thus I continued to live, vaguely confused until I went away from Pavia. Immediately it was discovered that the Fox, frisking and gloating because this had befallen, was promoted to the professorship by ruling of the Senate. But alas for mortal hopes! Not three or four times had he lectured before a disease attacked him which, according to the story which came to my ears, lasted about three months; and he died. He was utterly apprehensive because of his crime; for I understood afterwards that a certain one of his associates, as a material witness of his perfidious conduct, had been destined for a poison potion at a banquet.

Delfino died in that same year and Fioravanti shortly after. Nay, even in the machinations forged for my torment at Bologna, a similar fate overtook a like number of physicians, though somewhat later. Thus all who sought my life perished.

If, however, God had permitted me to be afflicted with so many calamities as a condition of the benefits which he has so unremittingly bestowed upon the human race, thus would my enemies have paid the reckoning. I had learned, moreover to keep at a distance from any chances of this sort, profiting by the fate of my uncle, Paolo, who had died of poisoning, and of my father who had twice drunk poison, and who, though he escaped, had lost his teeth.

Close upon the foregoing agitations followed I know not what a host of troubles. In the month of July I was forced to undertake a journey on account of a serious ailment with which my infant grandson was suffering at Pavia, while

I was at Milan. From the exertions of travel I fell ill of erysipelas, which settled in my face; this, followed by toothache, put me on the verge of having myself bled, when the approaching new moon caused me to refrain. From the day of the new moon, I deemed myself much improved, and so escaped both the peril of death and the pain of treatment. Subsequently a servant threatened murder in an argument over some money; I forestalled his act only by a few hours. In the wake of these troubles a long and painful attack of gout bore down upon me.

From the year 1572, I have suffered no slight risks of my life from lurking perils, since the streets of Rome are little short of a wilderness to me, and the behavior here so uncouth that many physicians more wary than I, and far more adroit in adjusting themselves to the customs, have found hereabout the origin of their deaths.

Accordingly, when I observed how I was protected more by Divine Providence than by any wit of my own, I ceased to exercise any further anxiety for my safety in dangers. Who does not now perceive that all these things have been, as it were, precursors of bliss about to be overtaken—a night-watch waiting for the dawn; just such a period as that which, in 1562, was followed by the professorship at Bologna, a post I held eight years, an honorable and useful work with a certain intermission from so many molestations and labor, and accompanied by a somewhat more pleasant existence.

Cardano did not mention that he cured the Scottish Archbishop of asthma by requiring that cats be banished from his presence. He may have thought that the cure was insufficiently dramatic.

17

Boole and Finite Differences

Both George Boole and finite differences deserve to have attention paid to them. Boole (1815–1864) never attended college and was self-educated in mathematics, but that did not stop Cork University from naming him professor of mathematics in 1849. His early death was the result of pneumonia, brought on by walking thorough rain to get to class and teaching in wet clothes and also, the story goes, by his wife, who held to the theory that the cure for a disease should resemble its cause and thus kept Boole wet.

Among his books are *An Investigation of the Laws of Thought*, from which we get Boolean algebra, and *Treatise on the Calculus of Finite Differences*. They have in common the goal of aiding thought by reducing it to symbol manipulation. It is easier to prove theorems in logic using Boolean algebra than it is to think the proofs through. Similarly, finite differences and the operator algebra that goes with it can make some things easy that otherwise might not be quite as easy. For example, here is how to sum consecutive squares:

$$x^2 + (x+1)^2 + (x+2)^2 + \cdots + (x+n-1)^2$$
$$= (1 + E + E^2 + \cdots + E^{n-1})x^2$$
$$= \frac{E^n - 1}{E - 1}x^2 = \frac{(1+\Delta)^n - 1}{\Delta}x^2$$
$$= \left(n + \binom{n}{2}\Delta + \binom{n}{3}\Delta^2 + \cdots\right)x^2$$
$$= nx^2 + \binom{n}{2}(2x+1) + \binom{n}{3} \cdot 2.$$

No thought needed! ($Ef(x) = f(x+1)$ and $\Delta f(x) = f(x+1) - f(x)$, so E and $\Delta + 1$ are the same operators.)

Δ is the discrete analogue of D, the differentiation operator. It was important—it is less so now than it used to be, computers being to blame—because
some information is inherently discrete. People die all at once, so insurance
actuaries have to deal with the discrete. Mortality tables have to be constructed
and smoothed, and finite differences were needed for that. One of the formulas
of the subject is named after Gauss and another for Bessel—even the giants of
mathematics made contributions.

What follows is a lecture, graceful and charming, by J. L. Synge (father
of Cathleen Morawetz, former president of the American Mathematical Society, and nephew of J. M. Synge, the author of *The Playboy of the Western
World*—who says talent doesn't run in families?) on the occasion of Boole
centenary in 1964.

George Boole and the
Calculus of Finite Differences

J. L. Synge (1969)

The world is so full of a number of things that we have to forget small things
in order to remember big things. A tombstone is not big enough to describe all
the merits of the deceased, and we must be satisfied with a simple slogan. Who
was Newton? The man who discovered the law of universal gravitation. Who
was Einstein? The man who discovered the theory of relativity, a theory which
no normal person need be expected to understand.

This tendency to describe a great man by a simple slogan was brought to
my attention when, some 35 years ago, the late Professor A. W. Conway and I
started to urge on the Academy the need for a collected edition of Hamilton's
papers. Who was Hamilton? The man who invented quaternions. That is quite
true, and I believe that quaternions will always be regarded as Hamilton's
greatest achievement. But of course he did other things of note in optics and
dynamics, and if you leave these out of account you get a very incomplete
picture of the man and his place in science.

Who was Boole? He was the man who applied algebraic symbolism to logical procedures, the inventor of Boolean algebra. That is the best simple slogan.

Boole's chief title to fame rests on his book published in 1854, entitled *An Investigation of the Laws of Thought on which are Founded the Mathematical Theories of Logic and Probabilities*. In 1954 the Academy celebrated this event by a symposium.

Now in 1964 we commemorate the centenary of Boole's death, and the President and Council of the Academy have done me the honor of inviting me to give a discourse on the occasion. I feel quite unworthy of the honor because I am not an expert in any phase of Boole's activities. However I am grateful for this compulsion to learn a little more about George Boole. I have thought it advisable to round out the picture of Boole by passing over his major work on the laws of thought and looking into another activity of his.

But first a few biographical facts. George Boole was born at Lincoln in 1815, the son of a tradesman of limited means. He became assistant master from the age of about 16 at schools in Doncaster and Waddington. He opened his own school in Lincoln, then became headmaster at Waddington and eventually, in 1840, moved this school to Lincoln. Meanwhile he had contributed his first published work, "Researches in the theory of analytical transformations" to the *Cambridge Mathematical Journal*. Further contributions to learned periodicals led to his appointment in 1849 as Professor of Mathematics in Queen's College, Cork. He died at Ballintemple in 1864, at the age of 49.

He wrote four books. The two first were *Mathematical Analysis of Logic* (1847) and the *Laws of Thought* (1854), referred to earlier. The other two were *Treatise on Differential Equations* (1859) and *Treatise on the Calculus of Finite Differences* (1860).

Throughout the whole of his work there runs a consistent vein of originality and daring. At least it must have seemed very original and daring at the time—thanks largely to Boole we have become thoroughly accustomed to it. It amounts essentially to the use of algebraic symbolism for the treatment of operators. Later I shall try to show what this means.

Leaving out then Boole's work on logic, we are left with differential equations and the calculus of finite differences. I decided to concentrate on the finite differences for several reasons.

One of those reasons is connected with sleep. Without a suitable stimulus, members of the Academy and distinguished visitors may fall asleep. There is of course no harm in that, but there is an unwritten law that the first duty of a lecturer is to keep his audience awake.

The best way to keep people awake is to irritate them or to challenge them to a contest of wits, as in a quiz program. So I give you a challenge.

Here is a sequence of six numbers

$$0 \quad 6 \quad 24 \quad 60 \quad 120 \quad 210$$

The question is this: What is the next number in the sequence, following those written down? The very posing of this question raises other questions, bordering on metaphysics. Is there a next number? You are at liberty to take a piece of chalk and write up any number and say that is the next number as far as you are concerned. In that way we might get as many numbers as there are members of this audience.

But perhaps there is some next number which is nicer in some way than any other? Some number that fits into a rhythmic pattern with the six numbers already written? Now we are getting somewhere. We are approaching mathematical sensitivity, for mathematicians are sensitive to rhythms of that sort. I venture to say that the members of this audience can be divided into four classes, one or more of such classes being perhaps empty:

(a) Those who pretty quickly feel intuitively out of a sense of rhythm what the next number must be.

(b) Those who remember that this discourse is concerned with finite differences and hence worry out the required number by the aid of finite differences.

(c) Those who cannot do this for themselves, but appreciate such a demonstration if it is worked out for them.

(d) Those who refuse to have anything to do with the matter, regarding it as sheer nonsense.

I shall now leave those six numbers in front of you, with the question "What is the seventh number?", and proceed to talk about George Boole and the calculus of finite differences. The audience can think about one thing or another, and if anyone falls asleep, there is material for interesting dreams.

What is the calculus of finite differences? I suggest that, to answer such a question, the most informative procedure is to give an example of a statement which belongs to the calculus of finite differences, the simpler the statement the better. In fact, any subject may be defined as a totality of meaningful statements, and one subject may be distinguished from another by giving statements belonging to one and not to another. Thus consider the statement that

> The costal cartilages are bars of hyaline cartilage which extend forwards from the anterior ends of the ribs and contribute very materially to the elasticity of the walls of the thorax.

That statement does not belong to the calculus of finite differences, but to anatomy. Likewise the statement

$$Dx^2 = 2x$$

belongs neither to anatomy nor to the calculus of finite differences; it belongs to, and is meaningful in, the differential calculus.

The beauty of the calculus of finite differences is that one can make a statement in it without involving oneself in technical terms like "hyaline cartilage" nor in the difficult ideas of limits which lie behind the differential calculus. In fact it is possible to exhibit statements which belong to the calculus of finite differences, statements transparent and obvious to a child of 10 years. Consider the sequence of numbers:

0   3   6   9   12   15   18.

Of this sequence one may say $\Delta u = 3$, meaning thereby that the *difference* between any member of the sequence and the next member is precisely 3. The statement $\Delta u = 3$ belongs, not to anatomy nor to the differential calculus, but to the calculus of finite differences. Consider now the sequence

0   1   4   9   16   25   36.

It is obvious that this sequence consists of the squares of the integers. That statement does not belong to the calculus of finite differences. But we can make statements about this sequence belonging to that subject. We prepare the following table

u	0	1	4	9	16	25	36
Δu	1	3	5	7	9	11	
$\Delta^2 u$	2	2	2	2	2		
$\Delta^3 u$	0	0	0	0			

The first line shows the given sequence, designated by the general symbol u. The next line shows the first differences, formed by writing under each number the difference between it and the next number. The third line shows the second differences, these being in fact the differences of the first differences. The fourth line consists of the third differences, which are the differences of the second differences. Thus, for the sequence which consists of the squares of the integers we may make the statement that the third differences are all zero, or symbolically $\Delta^3 u = 0$. This is a statement, not in anatomy nor in the differential calculus, but in the calculus of finite differences.

It would of course be entirely wrong to say that the calculus of finite differences consists entirely of statements transparent and obvious to a child of 10 years. But I should insist that it does contain such simple statements, and that is more than can be said about the differential calculus, which, as I have said involves the difficult idea of a limit, in other words, it involves the concept of infinity.

It is usual and indeed natural that the simple should precede the complex. Since the calculus of finite differences is simpler than the differential calculus, one might have expected it to be developed first. But that was not the case. The differential calculus was evolved by Newton and Leibniz, but the calculus of finite differences did not come until much later. Since the educational pattern follows the historical pattern, we find the differential calculus included as an essential part in standard mathematical curricula, whereas the calculus of finite differences is rather in the nature of a luxury except in connection with certain other mathematical disciplines. My own mathematical education followed the traditional pattern, and it was not until long after I graduated that I came to realize that I really ought to know something about the calculus of finite differences. I needed it, not for its own sake, but because I came across physical problems in which difference equations were involved. At this juncture I was fortunate to hear of, and to buy, a book on the subject written by Professor Charles Jordan of Budapest, translated into English in 1939 by Professor H. C. Carver of the University of Michigan. Professor Carver's introduction to this book is one of the reasons why I elected to speak about George Boole and the Calculus of Finite Differences. Professor Carver's authority is so much greater than any I could claim, that I am happy to stand aside and let him say what has to be said about Boole. Permit me then to quote a section from Professor Carver's introduction

> There is more than mere coincidence in the fact that the recent rapid growth in the theory and application of mathematical statistics has been accompanied by a revival in interest in the Calculus of Finite Differences. The reason for this phenomenon is clear: the student of mathematical statistics must now regard the finite calculus as just as important a tool and prerequisite as the infinitesimal calculus.
>
> To my mind, the progress that has been made to date in the development of the finite calculus has been marked and stimulated by the appearance of four outstanding texts. The first of these was the treatise by George Boole that appeared in 1860. I do not mean by this to underestimate the valuable contributions of earlier writers on this subject or to overlook the elaborate work of Lacroix. I merely wish to state that Boole was the first to present this subject in a form best suited to the needs of student and teacher.
>
> The second milestone was the remarkable book of Norlund that appeared in 1924. This book presented the first rigorous treatment of the subject, and was written from the point of view of the mathematician rather than the statistician. It was most opportune.

That ends the quotation. Professor Carver proceeds to name, as the other two outstanding texts, Steffensen's *Interpolation*, and the book by Jordan to which this is the introduction. It appears then that Boole's book held the field for over 60 years—from 1860 to 1924.

In the author's preface, Professor Jordan writes that his book is the result of nineteen year's lectures on the Calculus of Finite Differences, Probability, and Mathematical Statistics in the Budapest University of Technical and Economical Sciences, based on the venerable work of Stirling, Euler and Boole; he further mentions that he has utilized and further developed Boole's symbolic methods. The index lists Boole's first and second summation formulae, Boole's polynomials, and Boole's series.

If one were trying to give a brief introduction to the calculus of finite differences to students traditionally trained, one would, I am sure, try to establish links with the differential calculus, since that would be a domain familiar to such students. But on the present occasion that would not do, for I am proceeding on the naive assumption that the present audience is not acquainted with the differential calculus. On its own merits, then, let me try to explain the operational method which owes so much to George Boole. But one thing I would beg of you. Think of the thing in itself, and not as a tool for the enlargement of human comfort: it is indeed such a tool, essential in statistics, but please forget that, and play with the ideas as a child plays, for that is the true way of science.

I have to speak of three operators Δ, E and M (I use the notation of Jordan). An operator is something that does something to something when it is applied to that something. In this affair we have operators and operands, the operator is the surgeon and the operand is the patient. In our present work, the operand is a sequence of numbers, lying anaesthetized on the table. Let us, for purposes of illustration, take the sequence mentioned earlier:

$$
\begin{array}{llllll}
u & 0 & 6 & 24 & 60 & 120 & 210 \\
\Delta u & 6 & 18 & 36 & 60 & 90 \\
\Delta^2 u & 12 & 18 & 24 & 30
\end{array}
$$

Here I have written down the results of applying the operator Δ, as we did earlier, and then applying it for the second time. Now take the operator E, which means that we are to replace each member of the sequence by the next member:

$$
\begin{array}{llllll}
Eu & 6 & 24 & 60 & 120 & 210
\end{array}
$$

Finally, we take the operator M, which demands that we replace each term by the arithmetic mean of itself and the next term:

$$Mu \quad 3 \quad 15 \quad 42 \quad 90 \quad 165$$

We have then attached meanings to the operators Δ, E, M when applied to any sequence. Since these operations can be repeated, we attach meanings to *powers* of these operators Δ^2, Δ^3, ... , E^2, E^3, ... , M^2, M^3,.... What about the operator $(1 + \Delta)$? It has not been defined. At this juncture we need the daring courage of George Boole. Let us define this operator by using a familiar law of algebra as if the operator were a number:

$$(1 + \Delta)u = u + \Delta u.$$

This conveys a definite instruction: add to the original sequence the sequence of first differences: thus

$$(1 + \Delta)u = 6 \quad 24 \quad 60 \quad 120 \quad 210.$$

Compare this with the sequence Eu ; it is precisely the same. The operator $1 + \Delta$ is the same as the operator E. True, we have only verified this for a particular sequence as operand, but it is quite easy to see that it is true for all sequences. In fact, we have

$$E = 1 + \Delta, \quad M = 1 + \tfrac{1}{2}\,\Delta.$$

Remember that these are not equations connecting numbers: they connect operators. Many thrilling and perhaps dangerous suggestions enter at this point. They are thrilling because we are straining a mathematical symbolism beyond its original scope, and they are dangerous because we may find our selves talking nonsense.

Looking at the two equations last written, the mischievous idea comes to mind: eliminate Δ. This gives

$$2M - E = 1.$$

This means that, assuming we can proceed as in elementary algebra,

$$2Mu - Eu = u.$$

Let us check this:

$$
\begin{array}{rccccc}
2Mu = & 6 & 30 & 84 & 180 & 330 \\
Eu = & 6 & 24 & 60 & 120 & 210 \\
2Mu - Eu = & 0 & 6 & 24 & 60 & 120
\end{array}
$$

which is indeed the original sequence u.

How far can we go with this business? Quite a long way. Some kind of magic is at work.

We recall that Boole and Hamilton were contemporaries, Boole being 10 years younger and the two of them dying within a year of one another. The mysterious feature of Hamilton's quaternions lay in the non-commutative character of multiplication: $ab \neq ba$. We at once ask whether the operators Δ,

E, M commute when multiplied together. But does it mean anything to multiply them? What are we to understand by operators $E\Delta$ and ΔE for example? The instructions are perfectly clear, and we carry them out as follows:

$$
\begin{array}{llllll}
\Delta u = & 6 & 18 & 36 & 60 & 90 \\
E\Delta u = & 18 & 36 & 60 & 90 \\
Eu = & 6 & 24 & 60 & 120 & 210 \\
\Delta Eu = & 18 & 36 & 60 & 90
\end{array}
$$

In fact, $E\Delta u = \Delta Eu$, or $E\Delta = \Delta E$. The operators Δ and E do commute.

But let me clarify the train of thought. For purposes of illustration, I have been using a certain definite sequence. I have merely verified that Δ and E commute when applied to this sequence. Actually they commute when applied to any sequence. It is easy to prove, but I refrain.

In fact, the three operators Δ, E and M all commute with one another. This means that, in using them, one can relax and employ ordinary algebraic procedures.

We have been playing round with the sequence which I originally wrote down as a challenge. But so far I have not said what I think the next term should be. Let me now open up on this.

The trick is to write down the original sequence and its several differences:

$$
\begin{array}{llllllll}
u & 0 & 6 & 24 & 60 & 120 & 210 & [336] \\
\Delta u & 6 & 18 & 36 & 60 & 90 & [126] \\
\Delta^2 u & 12 & 18 & 24 & 30 & [36] \\
\Delta^3 u & 6 & 6 & 6 & [6] \\
\Delta^4 u & 0 & 0 & [0]
\end{array}
$$

The remarkable fact about the given sequence of numbers is that $\Delta^4 u = 0$. At least for the terms shown. If you are asked to continue the sequence, by supplying the 7th term, there is a psychological necessity certainly not a mathematical necessity to continue this rhythm. So we enter [0] in the last line. We are now committed to adding a definite term to each line, as shown.

So the answer is 336. Who would have thought it?

How does one make up little puzzles like this? Are there other sequences which give zero 4th differences? In other words, what sequences u satisfy the difference equation $\Delta^4 u = 0$?

The answer is that the nth term in the sequence is of the form

$$an^3 + bn^2 + cn + d$$

where a, b, c, d are any chosen numbers. I made up the above puzzle by taking $a = 1$, $b = 0$, $c = -1$, $d = 0$, so that the general term is $n^3 - n$.

If we pass beyond the very simple instances I have given, the calculus of finite differences involves some rather complicated formalism. But I would like to refer to a few formulae in order to put Boole into connection with two illustrious names from the eighteenth century—Bernoulli and Euler.

Let me remind you of the three operators Δ, M, E together with the operation of differentiation D. For any function $f(x)$ we have

$$\Delta f(x) = f(x + 1) - f(x)$$
$$Mf(x) = \tfrac{1}{2}[f(x + 1) + f(x)]$$
$$Ef(x) = f(x + 1)$$
$$Df(x) = f'(x)$$

We have as earlier

$$E = 1 + \Delta, \quad M = 1 + \tfrac{1}{2}\Delta$$

and by virtue of Taylor's series we have

$$E = e^D.$$

All four operators commute.

Following the notation of Jordan, I would like to exhibit the formulae connected with four sets of polynomials, viz. Bernoulli's first polynomials, Euler's polynomials, Bernoulli's second polynomials, and Boole's polynomials. In each case the equation on the left is essentially the defining condition, to within a constant, and that constant is supplied by the auxiliary condition on the right.

Bernoulli's first polynomials:

$$\Delta\varphi(x) = \frac{x^{n-1}}{(n-1)!} \qquad D\varphi(x) = \varphi_{n-1}(x);$$

$$\varphi_0(x) = 1, \ \varphi_1(x) = x - \frac{1}{2}, \ \varphi_2(x) = \frac{x^2}{2} - \frac{x}{2} + \frac{1}{12}, \ \varphi_3(x) = \frac{x^3}{3!} - \frac{1}{2}\frac{x^2}{2!} + \frac{1}{12}x + 0, \ldots$$

Euler's polynomials:

$$ME_n(x) = \frac{x^n}{n!} \qquad DE_n(x) = E_{n-1}(x);$$

$$E_0(x) = 1, \ E_1(x) = x - \frac{1}{2}, \ E_2(x) = \frac{x^2}{2!} - \frac{x}{2}, \ E_3(x) = \frac{x^3}{3!} - \frac{x^2}{2\cdot 2!} + \frac{1}{24}, \ldots$$

Bernoulli's second polynomials:

$$D\psi_n(x) = \frac{x}{n-1} \qquad \Delta\psi_n(x) = \psi_{n-1}(x);$$

$$\psi_0 = 1, \ \psi_1 = x + \frac{1}{2}, \ \psi_2 = \binom{x}{2} + \frac{1}{2}\binom{x}{1} - \frac{1}{12}, \ldots$$

Boole's polynomials:

$$M\zeta_n(x) = \binom{x}{n} \qquad \Delta\zeta_n(x) = \zeta_{n-1}(x);$$

$$\zeta_0(x) = 1, \zeta_1(x) = x - \frac{1}{2}, \zeta_2(x) = \binom{x}{2} - \frac{1}{2}\binom{x}{1} + \frac{1}{2^2}, \ldots$$

In each case the difference operator, and also M, refers to the increase of x by unity, n being unchanged. The symbol $\binom{x}{m}$ is the binomial coefficient

$$\binom{x}{m} = \frac{x(x-1)\cdots(x-m+1)}{m(m-1)\cdots 1}.$$

In each case the first few polynomials are written out.

Boole's *Treatise on the Calculus of Finite Differences* was published in 1860. Eight years after his death, in 1872 in fact, a second revised edition of the book was brought out under the editorship of John F. Moulton of Christ's College, Cambridge. This second edition was issued as a paperback by Dover Publications, New York, in 1960, and by Constable and Co. of London. The book is therefore easily available.

In his preface Moulton says that he has allowed himself considerable freedom as regards the form and arrangement of those parts where additions were necessary, but strictly adhered to the principle of inserting all that was contained in the first edition. I have not myself seen the first edition, accepting on trust that Moulton had not deviated too far from Boole. However last Saturday, in attempting to compose a conclusion to this discourse, I studied Moulton's preface, and I began to have some doubts. There is a certain difference between the atmospheres of Cork and Cambridge. It would be immodest for a Corkman to say "Where Cork leads, Cambridge follows!", but in 1872 this was true in respect of the subject-matter of Boole's book.

Boole was certainly not a dilettante, but he did his greatest work by virtue of a wide-reaching mind. It is therefore a little disquieting to find Moulton exposing in his preface a somewhat crabbed point of view. Moulton deplores the fact that the course of mathematical study at Cambridge puts a great premium on wide superficial reading. He writes: "But though this is at present the case, there is every reason to hope that it will not continue to be so." He looked forward to the time when students will aim at an exhaustive study of a few subjects in preference to a superficial acquaintance with the whole range of mathematical research. Well, Moulton was a good prophet. What he wished seems to have come to pass. The age of specialization is upon us. I have had the honor to serve for the past two years on a Mathematical Committee of the Royal Society and it seems to me that specialization in the various divisions of mathematics has now reached such an intensity that what one might call general mathematical conversation is no longer possible. There is of course everything to be said in favor of expert technical knowledge, but you have to pay the price for it—perhaps no more Booles.

Boole devotes one chapter in his treatise to the calculus of functions. Moulton retained it, but deplored it and also other material in the latter part of the book, because "it treats in a brief and necessarily imperfect manner subjects that had better be left to separate treatises." It is true that Boole was wandering a bit, because the calculus of functions is not a part of the calculus of finite differences but a generalization of it. For my part I think Boole did well to include it, for the simple reason that it is interesting. One of the exercises in this chapter is the derivation of the Gaussian law of error by means of a functional equation.

Mathematics is a very exhausting subject. It would be far too exhausting for anyone to pursue were it not for the invention of mathematical symbolism which might be described as canned thought. Someone has invented the symbolism and tested it, and once you have got confidence in the symbolism you can use it with ease. The intense cerebral activity of thought is replaced by the simple exercise of making marks on a piece of paper. Those who invent good symbolism in mathematics are like the inventors of roller skates or jet planes; by their aid you reach your destination much more quickly and with much less mental exhaustion.

In the matter of words, mathematics is atrocious: just think of negative numbers, imaginary numbers, irrational numbers, and so on. In the matter of symbolism it is better but far from perfect. About forty years ago I heard an American mathematician speaking about mathematical education—it was Prof. Hedrick, known by name to many as the translator of Goursat's book on Analysis. Hedrick thought that the most fundamental thing to get across to a student was the idea of a function. But what a word! We blush when someone refers to the functions of the body, and when a mathematician mentions a function the mind of the layman goes blank.

I shall not attempt to explain to those who do not know it what a function is. The usual symbol is $f(x)$. Boole would have written it fx. He regarded f as an operation applied to a variable number x, leading to a new variable number fx. The parentheses are confusing and unnecessary.

Boole's notation for a function is stimulating. It makes you wonder whether you can play with the symbolism. For example what about

$$(1 + f)x \qquad f^2 x \qquad f^{-1}x?$$

Do these symbols mean anything, and if so what? And what about

$$fgx \qquad gfx?$$

Are they equal? Of course not, as every modern physicist knows. For Boole's notation is actually accepted and used by those who need a compact and suggestive notation.

The time has come to bring to an end this brief and inadequate tribute to the memory of George Boole. We can now forget him until we reassemble in the year 2015 to celebrate the bi-centenary of his birth.

In his poem *Thus Spake Zarathrustra* Nietzsche said that the values by which men live are not changed by the noisy blustering people but by quiet men. In his quiet way George Boole established new values in mathematical thought.

There are perhaps several morals to be drawn from all this. First, that Cork is a good place to do mathematics in. Secondly, that really original thoughts are not the result of highly developed organizations but of original minds working in solitude. If there are any more morals to be drawn, you must draw them for yourselves.

The functional equation that gives rise to the Gaussian distribution is

$$f(x)f(y) = f\left(\sqrt{x^2 + y^2}\right)f(0),$$

whose solution (findable in a few lines: differentiate the equation with respect to x and with respect to y to see that $\dfrac{f'(x)}{xf(x)} = \dfrac{f'(y)}{yf(y)}$) is $f(x) = f(0)e^{-h^2 x^2}$.

J. L. Synge (1897–1995) was born and died in Dublin, though he spent 1930–1948 in North America, among other things heading the mathematics departments at Ohio State and Carnegie Tech. His lectures, as the example above shows, were clear and effective, as befits those of someone who saw them as combinations of "a theatrical performance, a religious sermon, and a circus."

18
Calculating Prodigies

The human race really is variable. There are people who look ordinary, even as you or I, but who are extraordinary—two, three, four, or more standard deviations away from the mean. People who can throw baseballs ninety-five miles an hour or improvise fugues on the organ may look *slightly* different from us, but not enough to startle. There are many mathematicians who are two or more orders of magnitude more talented than I, but this cannot be deduced from their appearance, or from their actions (when they are not doing mathematics, that is).

Every now and then there appear people who can do amazing things with numbers in their heads. They also are indistinguishable from the ordinary run of humanity. There is variability even within their group: some are above average in non-numerical intelligence while others would be classed as retarded. They do their marvels in many ways. Some see numbers, others hear them. (I once heard a gifted mental calculator ask that the air conditioning be turned off during one of his performances because it interfered with his ability to *hear* the numbers he was dealing with.)

Their minds are, to ordinary mortals, wondrously strange. What are we to make of the prodigy who knew 31,812 digits of π (a record that has since been surpassed) who "finds the numbers from the 2,901th to 3,000th places—81911979399520614196 and so on—particularly melodic. The series between 3,701 and 3,800—beginning 2322609729—is 'very jarring' "? We can only be amazed.

The calculators themselves do not think that they are strange. As one said, mental calculation is

> Easy to start with, and getting even easier as you go along. I would not call the green pastures of arithmetic an idiot's delight, but you need no

towering IQ to discover the key to numeric bliss. It is ideal relaxation for lazy people. In fact, once you have passed a critical threshold you don't even have to do the work yourself; a little man in your brain—or a tiny Turing machine—takes over and processes the numbers, then lays out the results for your inspection and approval.

And J. S. Bach said that anyone who would work as hard as he did would get just as far. Sure! People who run a mile in four minutes probably find that natural too.

Steven Smith's *The Great Mental Calculators* surveys how calculators do what they do, and gives minibiographies of some twenty of them. There follow two examples. They both show how unusual, and how utterly ordinary, calculating prodigies are.

The Great Mental Calculators

Steven B. Smith (1983)

Johann Martin Zacharias Dase

"He [Dase] is probably the most outstanding mental calculator of all peoples and all times."   —Dase (1849)

Information regarding the life of Johann Dase comes largely from his haphazard appearances in the voluminous correspondence between the mathematicians C. F. Gauss and H. C. Schumacher (1861).

Dase was born in Hamburg, June 23, 1824, the son of a distiller. According to a brief biographical sketch Dase (1849) himself wrote, he "developed through his own hard work, his inborn calculating ability and it has, it seems to him, been influenced only very slightly by his early schooling."

By the time he was 15, Dase was making appearances as a professional mental calculator. A year later he met Schulz von Strasnicky, a Viennese mathematician, who showed him how to compute "by the formula:[1]

$$\pi / 4 = \tan^{-1}(1/2) + \tan^{-1}(1/5) + \tan^{-1}(1/8).$$

---

[1] π is the ratio of the circumference of a circle to its diameter, that is $\pi = c/d$. It is an irrational number, which means that no matter how many decimal places it is computed to, it will never repeat.

Dase worked on the problem for two months, and came up with a result to 205 places, which was published in 1844 in *Crelle; Journal für Mathematik* (27:198). Twenty years earlier, William Rutherford of England had calculated pi to 208 places, but his result disagreed with Dase's from the 153rd place on. Dase was correct to 200 decimal places.

Sometime near his twentieth birthday Dase came to Altona, where Schumacher lived. He gave an exhibition of his talents but occasionally miscalculated considerably. He was especially fond of extracting fifth roots because, Schumacher remarked (Gauss and Schumacher 1861 3:382): "He had noticed that at the fifth power the units were the same as they were in the root."

Dase knew almost nothing of mathematics, even though Petersen, a prominent mathematician, had wasted a lot of effort in an attempt to teach him. Dase was at this time traveling about making money from exhibitions of his mental calculations. Schumacher was mystified by this (ibid.):

> Now he has fallen in with an oboist or corporal in the Hamburg Militia, with whom he travels, and lives from his exhibitions, although I can scarcely understand why people would give money to see him solve problems in his head. If there are reliable witnesses, that he can do it, then one gets nothing new from the exhibitions. All you see is a young man with a rather simple-looking face pronounce the right answer after a little while.

A year later Schumacher had the opportunity to test Dase personally. Dase made errors in every problem. A man who accompanied Dase attributed his poor performance to a bad headache and said that the great concentration required in his calculations frequently resulted in such headaches. But Schumacher remained doubtful in a letter dated August 8, 1845 (ibid. 3:32): "He smelled then somewhat of rum, probably accidently, since he is not supposed to drink anything but water."

Schumacher's only remarks on the problems submitted to Dase were that he gave him two six-digit numbers to multiply, which were done only partly mentally:

> He divided, I think, each number into two parts, in which the first contained the highest number and three zeros, the second the three lowest numbers. He calculated the four partial products in his head and wrote down each one with a pencil, which he then added together in his head. Perhaps he has been doing better feats now, since later in Berlin he took one thaler entrance fee, which one would scarcely pay for the feats he did for me.

A year and a half later, Dase was again in Altona, and this time Schumacher (ibid. 5:277) was much more favorably impressed. Dase was able to count at a glance the number of peas thrown on a table, and instantly added the spots on a group of dominoes (117). Schumacher reports that he was able to multiply and divide large numbers in his head, though no specific examples are given. Dase also succeeded in extracting the square root of a 100-digit number in his head in 52 minutes.

Dase was in the process of calculating the natural logarithms to seven places of the numbers from 1 to 1,005,000, of which two-thirds had already been completed. He expected to finish by Easter and was casting about for a publisher, which, Schumacher added, he would have trouble in finding.

By April 1847, Schumacher was describing Dase (ibid. 5:295) as a "strange calculating genius" (*sonderbare Rechengenie*). Dase had mentally multiplied two 20-digit numbers in 6 minutes, two 48-digit numbers in 40 minutes, and two 100-digit numbers in 8 3/4 hours. Apparently Schumacher himself did not witness these calculations, for he remarks: "The feat [of multiplying two 100-digit numbers] must have made the production rather boring." But apparently he did see a demonstration of Dase's root extracting ability, for he says: "Square roots of a 60-digit number he extracts in an unbelievably short time, though I did not note the time, and I cannot recall just how long it took."

Schumacher also informed Gauss that Dase was anxious to visit him so that Gauss could witness, and presumably comment favorably upon, his abilities. "So that he may blend the useful with the agreeable, however, he doesn't want to come until the students have come back to the lectures, so that he may give what he calls his production there."

In his reply (ibid. 5:296) Gauss was not at all encouraging. He pointed out that the preceding winter a Dane, who gave a similar demonstration, failed even to cover his costs. "In fact, the cost of the trip, the cost of the stay, the cost of the police permit (if, in fact, they will give him one; another traveling performer, Holtey, who wanted to give an exhibition of oratory here was denied papers), the rent, heating, and lighting of a hall, etc., would probably cost more than the admission price would realize."

In any event Gauss was not much impressed with what he had heard of Dase's calculating prowess.

> From what I have learned through letters or published papers that have come to my attention, there is little testimony for any outstanding ability for calculation. One must distinguish two things here: an extraordinary memory for numbers and true calculating ability. These are, in fact, two completely separate qualities that may be connected, but are not always.

One person might have a very good memory for numbers without being able to calculate well .... On the other hand, one can have a superior ability to calculate without having an unusually strong memory for numbers. The latter, Herr Dase has without doubt to an eminent degree. I confess, however, that I can attribute but little worth to it. Calculating ability can only be assessed as to whether someone does as well or better on paper than another person. Whether this is the case with Herr Dase I don't know; only when he tries to multiply two numbers, each one of 100 digits together in his head and it takes him 8 3/4 hours, that is in the final analysis a crazy waste of time, since a somewhat experienced calculator can do the same on paper in a much shorter time; he could do it in less than half the time ... [in any case] has the correctness of his calculations been tested?

Certainly a pertinent question when such immense numbers are involved, but a question to which Schumacher made no direct reply.

By this time Schumacher seems to have become something of an apologist for Dase. In his April 12 reply to Gauss (ibid. 5:300) he reported that Dase also was excellent when computing on paper.

In another letter the following day (ibid. 5:301–2) Schumacher wrote that there was no dissuading Dase from his intention to go to Göttingen, for he wanted to learn from Gauss what really important calculations he should carry out. In this letter, Schumacher also gives some concrete examples of Dase's numerical memory and his calculating skill. Schumacher first wrote down the number 713,592,853,746. Dase complained that the numbers were written too small; he wasn't sure whether the eighth digit was a five or an eight. Schumacher then wrote 935,173,853,927. This Dase glanced at "for about a second" and was then able to repeat it backward and forward. Dase then offered to multiply this number by a multiple of Schumacher's choice. He selected 7, which was, of course, child's play for Dase, who immediately replied 6,546,216,977,489. As he was leaving, Schumacher asked him whether he remembered these numbers and he promptly strung them together as one number of 25 places and repeated them forward and backward.

Schumacher also had Dase multiply 49,735,827 by 98,536,474. This he did in the ordinary way on paper in 1 minute and 7 seconds: 4,900,793,024,053,998. Dase complained, however, that the necessity to write the number down slowed him up and that he could calculate faster in his head. Schumacher then gave him the numbers 79,532,853 and 93,758,479, which Dase multiplied in his head in 54 seconds: 7,456,879,327,810,587. Dase also counted, at a glance, the number of letters in two different lines of print in books (47 and 63).

Gauss was plainly displeased by Dase's projected visit. In his letter of April 16 (ibid. 5:303), he again pointed out the likelihood of financial failure, and that, furthermore, Dase would probably also fail in his other goal. "I have thought about the problem, but I can't recommend any job that would be proportionate to his abilities ...."

Gauss did acknowledge, however, that he personally would like to see Burckhardt's table of factors, completed in published form up to 3,000,000, continued. The problems were that no great calculating ability was needed to do it, and the market for such a work was likely to be very small, so that Dase could expect little reward for his labor. In fact, there was a continuation up to 6,000,000 in manuscript in Berlin, but nobody was willing to bear the costs of publishing it. As far as Dase's being of any use to Gauss himself, he wrote: "I, in my own life, have carried out very large calculations, very many of them, and occasionally I have used outside help; but I fail to remember any case in which purely mechanical calculating help, no matter how high the sum or big the problem, was of any particular use to me."

Gauss could recall a couple of cases in which calculators had been hired (a Professor Petzval had hired 23, paying them very well—a "Kaiserly wage"— to draw up a table for optical lenses), but he knew of no such opportunities at the time.

In the end Dase did not go to Göttingen. Not until two years later, in a letter from Schumacher to Gauss in May (ibid. 6:28), does Dase reappear in their correspondence. Dase had written to Schumacher from Vienna of his intention to continue the factor and prime number table up to ten million. For this the support of the Austrian Academy was needed, but before making a decision, the academy required Gauss's judgment. Schumacher remarked: "It seems to me, if the work is in itself useful, to be a completely appropriate use of such a calculating machine. There is nothing more to it than that; there is no *divinae particula aurae* in him."

Another year passed and Dase again went to Altona (from Vienna). His table of natural logarithms had been printed. He had three copies out of a promised 500, but the rest were in Vienna and would remain there until he came up with the money to pay for the paper. Dase was on his way to England to earn enough by performing to get his books out of hock. Schumacher laments: "It seems to me that the Austrian regime should have offered the poor devil the 326 fl. he needed."

In September of 1850 Dase was still hoping to go to Göttingen to give a "production." Again Gauss (ibid. 6:112) was discouraging:

... I can only repeat that exhibits of his calculating ability before this public, as far as I can judge of their taste, would find only very little success, and would barely pay the cost of the trip, and that here I can scarcely foresee any occasion where his other talents would be of any use to him as far as making money is concerned.

Gauss did, however, in December of 1850, send Dase a lengthy letter outlining the history of tables of factors—quoted in the foreword to Dase's (1862) factor tables—and suggested:

in my opinion the most immediately desirable task would be the working out of the four million from 6,000,000 to 10,000,000, naturally not to the exclusion of an eventual even longer continuation, given the ability to carry it out. You, yourself, possess several of the needful qualities to do so in outstanding measure: an excellent proficiency and sure grasp of arithmetical operations and, as you have already shown in several cases, an indestructible determination and patience. Thus, should you find yourself in a position, through the sponsorship of the well-to-do, the scientific aspirations of the well-disposed citizens of your country, or by any other means, to undertake such a task, it would be well received by the friends of arithmetic.

As a result of Gauss's recommendation Dase was able to find support from "several promoters of science in Hamburg," somewhat less than a year later. He died ten years later in 1861, at the age of 37. He had, by that time, completed the entire seventh million and all but a small portion of the eighth; he had also finished a considerable part of the ninth and tenth millions. The seventh million was published posthumously, in the year of his death.

Dase was a man determined to leave some mark of his passing on the world, as his search for a really important calculation to carry out attests—a calculation that would be valued by a universally recognized genius like Gauss. Ironically, the tables over which he labored for so many years and which bore his name are a trivial exercise for a modern electronic computer.

Wim Klein

In Marseille on a Sunday afternoon, there was a chap who was eating flames, there was a lady who told the future, there was a chap with a monkey, and I was extracting cube roots. Ho, ho. I'd like to go back and do it again. It was fun. But I did not need to do it. I just got so fascinated with this vagabond life.

—Wim Klein, reminiscing about the postwar period[2]

_____

[2] Unless otherwise attributed, information in this chapter is based on personal interviews with Wim Klein.

Wim Klein (also sometimes referred to as Willem or William Klein) was born in Amsterdam on December 4, 1912—a Wednesday, as he is quick to tell you.

Klein's interest in calculation began at age 8, when he discovered factoring. "At school we had to factor numbers up to 500. Then I continued on to 10,000, 15,000, 20,000, 25,000. As you got so often the same combinations, it is logical that if you know that 2,537 is 43 times 59, and you're doing a little show for the godmother of a neighbor celebrating her eighteenth birthday, and they ask you for 43 times 59, you recognize straightaway 2,537."

Although Klein never set about to learn the multiplication table up to 100 by 100, he gradually acquired it from repeatedly encountering the same combinations. He contrasts this sort of memory, which comes about unbidden simply as a matter of repeated exposure to certain material, to deliberately committing something to memory; the latter he calls "mechanical memory." "The multiplication and the squares up to 1,000, I just took as a game. Learning the logarithm table by heart up to 150 is memorizing, but in terms of multiplication I never memorized. It came from the experience I got by factoring—I got often that 2,537 is 43 times 59, and 5,074 is the double, and 7,611 three times the number so you recognize it."

By exposure he has learned the multiplication tables up to 100 by 100, the squares of integers up to 1,000, the cubes of numbers up to 100, and roughly all prime numbers below 10,000. He deliberately committed to memory the decimal logarithms to five places of the first 150 integers. He also knows "some other small things like the first 32 powers of 2, the first 20 powers of three and so on; some logarithms base e; a lot of history; and I also learned by heart the date of birth and death of about 150 composers."

Klein's older brother, Leo, was also an exceptional mental calculator, but Wim was the moving force behind the brothers' interest. "Leo was a little infected by me. Because I did it, he also had to do it."

The brothers were, however, altogether different in their methods—Leo's memory was visual, while Wim's is auditory. Leo did not share Wim's fascination for factoring, nor did he care to go beyond three digits by three digits in multiplication.

To illustrate the difference between his methods and Leo's, Klein asked me to call out two three-digit numbers. I chose 426 and 843. He muttered in Dutch and after a few seconds said: "359,118." He then took a piece of paper and wrote:

$$4.26$$
$$\underline{8.43}$$

$$12.78$$
$$170.40$$
$$\underline{3408.00}$$
$$3591.18$$

"My brother would say: 'There's twelve dollars and seventy-eight cents plus a hundred and seventy dollars and forty cents plus three thousand four hundred and eight dollars—three thousand five hundred ninety-one dollars and eighteen cents, which he had to translate into normal pronunciation as three hundred fifty-nine thousand one hundred eighteen.

"But I say, hey, 426 divided by 6 is 71, and 843 times 6 is 5,058; 5 times 71 is 355 [in this case thousand] and 58 times 71, by experience, without calculating it, is 4,118. You see the difference? If you had taken 427 I should have done quite differently: I should have said '427 divided by 7 is 61 and 843 times 7 is 5,901; 61 times 5,901 is 359,961.' You see how it helps using factorization as much as possible? There is a keyhole in a hotel and I've got a key ring with 500 keys. What I have to do is to pick the right key for that slot. For every problem I have to think straightaway which is the best way—just like a flash."

The difference between the brothers' memory showed up in other areas as well. Leo would come into a town, buy a map, scan it briefly, and know his way about. Klein recalls: "He would say to a waiter in a pub, 'I have to go there.' The chap would say, 'Well, you go like this.' My brother says, 'And if you go like this, is it not shorter?' The waiter says, 'Yes, sir.' "

When the brothers were children and took the streetcar to school, each conductor wore an identification number on his collar. Years later they saw an elderly man selling tickets on the streetcar. Leo asked Wim, "Do you remember him?"

"No."

"In the old days his collar number was 683."

Leo started to talk to the man and after a while he asked: "How many years have you worked on the tramway?"

"Oh, since ..."

"Yes, and years ago you were on streetcar number 60."

"Yes."

"And your collar number was 683."

The old man looked at him: "You damn ... That is correct."

Klein says: "There were hundreds of examples like this. Funny, eh? This visualizing, I cannot do it at all."

Klein's father wanted a successor to his medical practice, so Wim very reluctantly undertook the study of medicine, even though he was already dreaming of show business. In spite of all the time he was forced to devote to studies, he managed to give little shows now and then. He finished his theoretical studies "after a hell of a lot of trouble," but before he could finish medical school his father died, and Wim abruptly halted his studies. "The old man died in 1937. He was always very strict on the penny with my brother and me. After he died, we got a little inheritance. So the brothers took full profit of life. Then the war came. It became very tough, of course, as you may guess. I lost my brother in the war, also killed like hundreds of thousands of others."

For two years Klein worked in a Jewish hospital, as other hospitals were forbidden to Jews. Then the Germans started to take people from the hospitals to camps and Klein was forced into hiding. Leo was not so fortunate; he was sent to a camp in Germany from which he never returned. It is not a period Klein cares to recall. "I had to hide. Some people took care of me. Just say it was like the case of Anne Frank. That is sufficient."

After the war Klein had to find some way to make a living, since the Germans had confiscated all his money. "Everybody needed show business at that moment, after those awful years."

Klein's first postwar role as a professional calculator was a nonspeaking one. He was decked out as a sort of Indian fakir with turban and a false beard. His partner did all the talking while Klein chalked up the answers.

But the theatrical agency was dissatisfied: "They told me, 'Listen, this act with the beard stinks. Your presentation is vulgar. We will get you in contact with one of the best announcers in the country, and you will travel about with a group of excellent artists.' And then it developed as a really nice act. But then, as Wim Klein was too cheap, they came up with 'Pascal.' In Holland they don't know Wim Klein, they know only Pascal. Why Pascal? It is a French name, but Pascal was the inventor of the calculating machine. When Pascal came to France, where nine out of ten people call themselves Pascal, and my French with my Dutch accent was not good enough, so there I became Wim Klein."[3]

In France and parts of Belgium, Klein had to do his act in his "poor, bad, school French." (He later learned to speak fluent, if somewhat confusing, French, German, and English.) "There was an expression, which in Dutch was a normal expression, so I thought in French it should be the same, so I translated it literally from Dutch into French and it meant something quite different. Nobody told me, but the audience burst out in laughter the whole bloody week. They did not want to tell me what was wrong with it, and only the last day, they told me it meant something like 'How often in the week do you do it?' or something like that. Silly."

---

[3] Klein is also sometimes known as "Willie Wortel" (Wortel being the Dutch word for root) from his ability to extract roots of large numbers.

During the postwar period he appeared in France and Belgium, and began doing radio broadcasts in Holland. While in Brussels in 1949, Klein was down on his luck. "I spent all my bloody money. Then I met some friends and one said: 'I play the guitar and he plays the accordion. You have a blackboard. We make some music and you do some sums.' We went to nightclubs, little pubs. First they made the music, then I came on with my sums, and then the guitarist went around with the hat.

"After a while we decided to go to Paris. We went to the Champs Elysees where people sit outside, and started to perform, but the police came and said: 'Shut up, you bloody beggars.'

"My friends went back to Holland, and after that I joined a little circus. But we went broke. A chap said, 'Why don't you just set up your act at the subway entrance? Lots of people do it.' And yes, it worked perfectly, but there were two enemies—the rain and the police."

"Some cops would say 'We're here this week, but next week some bastards are coming, so next week go somewhere else.' Or a Dutchman would say, 'Hey, can you fix me up with a nice girl?' 'Yep.' 'Emanuel, I've got a guy for you.' 'Here, my dear, this is for you. A tip.' Oh, what a great time."

But as Klein had no work permit, the French authorities finally kicked him out of the country. At the train he met a Belgian who had seen him perform several times in the Place Pigalle. The man was living in Mons near the French border. He said that he had contacts in Belgium through which they could organize a lecture tour in the schools. "They put me up in a little pension, and they bought me some decent clothes. After three months time I was out of debt."

In 1952 Klein got a job at the Mathematisch Centrum in Amsterdam, where he did various sorts of numerical calculations. "Computers—they didn't exist, or nearly not. I sat in a room with these five heavy-reformist girls [members of the Netherlands Reformed Church], always talking about God and the clergy. I would say 'Good God' and 'God damn it,' so they went to the boss and said, 'Klein is swearing like a docker.' He told them, 'Don't quarrel, let him swear.' They said, 'Yes, but …' So he called me in and said, 'Listen, Klein, quiet down. I know they are idiots, but try to do better.' 'I try, I try very hard, but you know … ' "

In 1952 Klein began seriously lecturing in schools; again it came about by accident. Whenever important people visited the Mathematisch Centrum, Klein was called upon to give a demonstration, "not as a human computer, but as a human attraction." A French professor from UNESCO saw him there and asked him if he would come to Paris and give a lecture to the Department of Mathematics at the Sorbonne. The planned fifty-minute appearance stretched to two hours. As a result of contacts made there, Klein was able to obtain permission to give lectures in grammar schools throughout France. "So I wrote to the Mathematisch Centrum and asked for leave of two months or so. They replied, 'Wim, I'm afraid that this means the end—that you will stay longer than two years. But I'm sure you will have tremendous success.' "

In 1954 Klein met the New Zealand mathematician and calculating prodigy, Alexander Craig Aitken, at a mathematical conference in Amsterdam. Later in that year they appeared together on a BBC program. Klein recalls: "He was a lovely man. When Britain had not yet the decimal system, I used to do problems like multiplying £3, 7 shillings, 8 pence, ha'penny by 29. When I asked Aitken about such problems, he said: 'Oh, I've got enough trouble when I have to fill out my income tax form.' "

In 1955 Klein toured for nine months as one of the attractions of the "MIR-ACLES OF THE MUSIC HALL, Starring Some of the Most Unusual People Ever Seen." He was billed as "the man with the £10,000 brain." The cast included "the Dare-Devil Denglaros on their Racing Motor Bikes; the Amazing Devero, Escape from a Real Guillotine; Ladd West, World's Fantastic Aerial Contortionist; Rondart, World's only Dart Blower; the Roller Skating Jeretz from Geneva; Reggie 'yer see' Dennis, Britain's new radio Comedian; the Incomparable Mime Star, Danny O'Dare; and Personal Appearance of The Man Who Was Buried Alive—already seen by 4,719,329 people in 8 years tour."

Klein reflected on some of his fellow performers: "This chap was a real pig. He always escaped, but we all hoped that he wouldn't. Danny O'Dare, the Indian Rubberman, was just a poor devil. With the Motodevils, every second word was a swear word. You could not speak decently with those people.

"Tommy Jacobson, the armless wonder—the first thing he did was to take a rifle and shoot, then he played the piano with his feet, then the master of ceremonies asked someone to come on the stage for a shave. He told the chap to sit down and said, 'Tommy, not as much blood as yesterday.' So the poor chap turned his head so, and Tommy took a big knife and ... *sccrrr*.

"Once Tommy was standing on the platform of a London bus and a chap tried to grab hold of him to pull himself on the moving bus. He grabbed Tommy's raincoat sleeve, but there was nothing in it, so he tumbled into the street clutching a raincoat."

The Miracles of the Music Hall gave two shows a night. After doing his act for the first house, a performer was off until it was time to perform in the second show. Between shows one night, Klein dropped into a pub next door. People from the audience of the first show called him over, started buying him drinks and asking him to do calculations. By the time Klein finally went running back to the theater, the second show was almost over. Someone said to him, "Wim, be careful. The announcer is already in a bad mood."

The announcer, who was also the manager, called out, "Wim Klein, there he is, here comes Wim Klein." Klein says, "I was struggling up, you know. He still didn't notice, but then he caught a funny smell of gin. He came really close and said, 'You bloody Peruvian Chinese teapot, you. If you don't finish your act properly, I'll kick you out straightaway.' I said, 'Yeah, yeah, I've just been kidding.' So I made it, more or less, and after the show he said, 'Listen, I also like to drink, but never do it between the two houses. You promise?' 'Yes.' 'Come on, let's have a drink together.' So we both got pissed when the show was over."

After Miracles of the Music Hall, Klein returned to touring schools, first in France and then in Switzerland. But by 1957 he decided he wanted to settle down, so he returned to Amsterdam to work at the Mathematisch Centrum.

During the summer of 1958 he arranged a two-week tour of Swiss schools. The giant research complex of CERN (European Organization for Nuclear Research) is located near Geneva. Klein was mistakenly under the impression that some of the work of the Mathematisch Centrum was done for CERN, so he decided to telephone CERN while he was in Geneva. He was told to "just pop over." He was introduced to a Dutch physicist, C. J. Bakker. Klein recalled: "We talked and he said, 'Would you like to work in Geneva?' I said, 'That's for other people to say. Is there a possibility for me to get a job here?' And then Professor Bakker said, 'Listen, Wim, I cannot decide; I'm only the director general here, you see.' So I said, 'Not so bad.' Naturally, I had a feeling everything would be all right."

An arrangement was made with the Mathematisch Centrum for a three-month leave of absence. After four weeks, CERN asked Klein to stay on permanently. "These three months became eighteen years. That's the CERN story."

In the early days Klein was in considerable demand at CERN. "Computers were not very well developed, and the physicists did not yet program them themselves. From '58 to '65 it was all right for me. And then, it went down, because young physicists did their own programming, and so they did not need me as much as before. But the idea to kick me out never came, because of public relations. Very often when physicists came and they could not see the machine, someone would say, 'Hey, Wim, do something for them.' "

Jeremy Bernstein described an encounter with Klein (1963:20)

In the summer of 1961, I had an opportunity to work with Mr. William Klein, a programmer and numerical analyst for CERN ..., in Geneva, who must be one of the fastest human computers who has ever lived. I was spending the summer doing physics at CERN and had been working with a friend on a problem. After a week or so, we produced an algebraic formula that seemed admirable to us in many respects, and we wanted to evaluate it. CERN has a large Ferranti Mercury computer, and since at the time neither of us knew anything about programming, we asked for help. Enter Mr. Klein. Mr. Klein is a short, kindly, energetic-looking man in his forties. He is of Dutch origin. He looked at our formula for a few seconds, muttering to himself in Dutch, and then gave us numerical estimates for several of the more complex parts of it. Doing this, he said, helped set up the program for the computer in the most efficient way. I had heard about Mr. Klein's almost incredible ability, and I asked him whether he had considered evaluating our whole expression in his head. He told me that it would involve much too much work and that he was quite glad to turn the job over to the machine. Watching Mr. Klein at work made a deep impression on me ...

By the mid-seventies, Klein was growing weary of CERN. In 1975 Amsterdam was celebrating its 700th anniversary, and Klein visited there some eight times. The next year he again visited Amsterdam several times. He decided to retire. "It's too monotonous—18 years. It was a golden cage, but I prefer silver freedom. So at the end I retired one year before I was 65."

In June 1974, shortly before his departure from CERN, Klein became intrigued with the problem of extracting integer roots of large numbers. The 1974 edition of the *Guinness Book of World Records* reported that Herbert B. de Grote of Mexico City had extracted the 13th root of a hundred-digit number in 23 minutes.

Klein says: "What is the use of extracting the 13th root of 100 digits? 'Must be a bloody idiot' you say. No. It puts you in the *Guinness Book*, of course.

"I never came on the idea until I got this notice about this man in Mexico. I thought, hey, how interesting. I should have thought of that. First I had to find out how to tackle the problem. Then I needed material—I needed numbers raised to the wanted power. So they wrote a multiprecision program on the computer. And I was practicing like hell, like hell, like hell. Once you know the system for the first one, you have to learn another series of numbers by heart for the next one."

By October 8, 1974, Klein succeeded in extracting the 23rd root of a 200-digit number in 18 minutes, 7 seconds, and on March 5, 1975, in Lyon, he reduced the time to 10 minutes, 32 seconds.

Later, Klein went on to extract a variety of roots: the 19th root of 133 digits (1 m. 43 sec.), the seventh root of 63 digits (8 m. 27 sec.), the 73d root of 500 digits (2 m. 9 sec.).

As explained in chapter 13, the difficulty of extracting integer roots of large numbers depends on the number of digits in the root—the size of the power is immaterial. The *Guinness Book of World Records* now accepts the extraction of the 13th root of a 100-digit number as a fair test; records now hinge upon improving the time required.

Klein has continually improved his times for extracting such roots. In Providence, Rhode Island, in September 1979, he achieved a time of 3 minutes 25 seconds; then in Paris, November 1979, 3 minutes 6 seconds; Leiden, March 1980, 2 minutes 45 seconds; London (BBC), May 1980, 2 minutes 9 seconds; Berlin, November 10, 1980, 2 minutes 8 seconds. Finally, on November 13, 1980, he got below two minutes—1 minute 56 seconds. And on April 7, 1981, at the National Laboratory for High Energy Physics, Tsukuba, Japan, he established a new record of 1 minute and 28.8 seconds. With this he is fairly well satisfied. He plans now to attempt to split up the four- or five-digit numbers as the sum of four squares within one minute.

Klein describes much of his calculating as "semimental," in that he has the problem in view while solving it (thus obviating the necessity of memorizing it) and because he often writes down parts of the answer as he calculates them, before the entire answer has been found; this means that he does not need to keep the entire answer in mind before announcing it. For example, in multiplying he uses cross multiplication, which allows him to write down the digits from right to left as they are obtained. After multiplying two eight-digit numbers for me "semimentally," Klein remarked, "Some people say to write down results as you go is not fair. You have to do it all in your head. But it takes five times as long and the audience will say, 'Forget it.' "

He asked me for two five-digit numbers and I gave him 57,825 and 13,489. In 44 seconds he multiplied these together mentally, without any intermediate results. He then repeated the experiment, writing down the answer as he calculated it—the time required was 14 seconds. "The first is more scientific, if you want. It is the real thing, but it's not what the people want."

Klein's passion, apart from numbers, is music. He is particularly fond of jazz and classical music. He says, "In New York every night I went to Jimmy Ryan's Jazz Club. I also did that when I was in New York two years ago. So when I popped in this time, they said, 'Hey, Flying Dutchman, how are you? Have a drink. What shall we play for you?' " (Here Klein gave an excellent imitation of a trombone playing *Ain't Misbehaving*.)

"I play no instruments, pity enough. I've got about 600 LPs. Not very much, but ..."

In spite of the justifiable pride Klein takes in his calculating ability, there is a passionate honesty in him as well. When I suggested that he may be the world's greatest mental calculator, he replied, "I'm not the world's greatest calculator. Perhaps the world's fastest calculator." In any case, Klein is surely one of the best mental calculators in history.

Steven B. Smith has taught linguistics at UCLA and the University of California at Riverside.

19

James Smith, Circle-Squarer

Mathematical cranks—those people who trisect the angle, square the circle, prove Fermat's theorem, show that the real numbers are countable, and so on— we always have with us. For a survey of the field, I will immodestly mention my *Mathematical Cranks* (Mathematical Association of America, 1992).

Most crank literature is all business: pages and pages of mathematics (or pseudomathemtics) intended to convince the reader that the author has indeed shown that there are infinitely many twin primes, that Goldbach's conjecture is true, or whatever it is that the crank is aiming at. It is not often that a crank describes his experiences. There follows an example.

Augustus De Morgan, in his *Budget of Paradoxes*, devotes quite a bit of space to James Smith, a circle-squarer who maintained that $\pi = 3.125$. De Morgan found that his method of arriving at this value was to assume that it was correct and then show that, with that assumption, calculations using $\pi = 3.14159265...$ were not correct. Try as he might, neither De Morgan nor anyone else was able to make him see reason. Smith had the money to put his views in a 200-page book, *The Quadrature of the Circle*, whose subtitle is "Correspondence between an Eminent Mathematician and James Smith, Esq. (Member of the Mersey Docks and Harbour Board)"

It was published in London by Simpkin, Marshall & Co. and in Edinburgh by Oliver and Boyd in 1861. Most of the space in it is given over to Mr. Smith's letters, filled with interminable arithmetical calculations, with a small fraction for the Eminent Mathematician's short replies, which included such things as

> Can you seriously suppose that Newton, La Place, Des Cartes, &c., &c., and all the wise and learned philosophers who ever lived, were so grossly ignorant as to deceive themselves, on a matter which is suscep-

tible of the most distinct proof, and may be verified by actual oracular inspection by the humblest individual?

My letters to you, were written in the sincere conviction that I was writing to one earnestly engaged in the search after truth, and my observations were confined to the pointing out to him, how he might convince himself that he was altogether wrong. My letters were not intended for publication, and I protest against their being published, for I do not wish to be gibbeted to the world as having been foolish enough to enter upon, what I feel now to have been, a ridiculous enterprise.

As the E. M. should have known, appeals to authority are useless with circle-squarers, since they think that they have the truth no matter what anyone else may think or say. He should also have known that Smith was not searching after truth, but only seeking agreement, and nothing else would do.

The library of University College London has a collection of Smith's letters to the editor of the *Athenaeum* (cranks will write to anyone who will reply, and to many who will not) arguing against De Morgan. Here is what the library has to say about Smith:

James Smith was born in Liverpool on 26 March 1805, the son of Joshua Smith. He entered a merchant's office at an early age, and, after remaining there seventeen years, he started his own business, retiring in 1855. He studied geometry and mathematics for practical purposes, and made some mechanical experiments with a view to facilitating mining operations. He became interested in the problem of squaring the circle, and in 1859 he published a work entitled *The Problem of Squaring the Circle Solved*, which was followed in 1861 by *The Quadrature of the Circle: Correspondence between an Eminent Mathematician and J. Smith, Esq.* This was ridiculed in the *Athenaeum*, and Smith replied in a letter which was inserted as an advertisement. From this time the establishment of his theory became the central interest of his life, and he bombarded the Royal Society and most of the mathematicians of the day with many letters and pamphlets on the subject. Augustus De Morgan was selected as his particular victim on account of certain reflections he had cast on him in the *Athenaeum*. Smith was not content to claim that he was able graphically to construct a square equal in area to a given circle, but boldly lay down the proposition that the diameter of a circle was to the circumference in the exact proportion of 1 to 3.125. In ordinary business matters, however, he was shrewd and capable. He was nominated by the Board of Trade to a seat on the Liverpool local marine board, and was a member of the Mersey docks and harbour board. He died at his residence, Barkeley House, Seaforth, near Liverpool, in March 1872.

There follows Smith's introduction to his book.

The Quadrature of the Circle

James Smith (1861)

Introduction

"No amount of attestation of innumerable and honest witnesses, would ever convince any one versed in mathematical and mechanical science, that a person had squared the circle, or discovered perpetual motion."
—*Baden Powell, in Essays and Reviews, 8th. Ed., page 141.*

"The Quadrature of the Circle" is a problem which it has long been pronounced an impossibility to solve, on the authority of names, of such high standing in the scientific world both in ancient and modern times, that latterly, every man who has entertained the idea of attempting its solution has been regarded as a wild enthusiast. "You may as well attempt to square the circle," has passed into a proverb, which is as familiar to the peasant as to the philosopher, and the former, would probably with just as little hesitation as the latter, arrive at the conclusion, that the man who could imagine he had discovered its solution, was in a state of mind which hardly fitted him to be entrusted with the care of his own person. And yet, the solution of the problem is extremely simple after all. It would almost appear as if its very simplicity had been the grand obstacle which had hitherto stood in the way of its discovery.

"The British Association for the Advancement of Science" may assume infallibility, and authoritatively proclaim that the solution of the problem is impossible; and may consequently decline to permit the consideration of the subject to be introduced into their deliberations. Other learned Associations may add the weight of their authority by endorsing such a course of procedure. The Astronomer Royal of England *may make a display of his contempt* for any one who should venture to address him as an "authority" on the question. And yet, they must all ultimately succumb to the force of truth, however humiliating it may be to their pride to submit to the infliction.

The following correspondence arose out of a pamphlet which I published about the period of the last meeting of the British Association at Oxford. At that meeting I distributed about 500 copies of the pamphlet among the *Savans* there assembled, and at the same time, I forwarded a copy to each member of the two Houses of Parliament. In this way I distributed upwards of 1500 copies, among the most learned and best educated men in this country; and I did so, in the hope that some one of them would take up the consideration of this important question from my point of view, and afford me the opportunity of a candid and careful discussion of it. In this hope I have not been altogether disappointed.

My correspondent claims, and justly claims, to be an "authority" on this great question, having obtained the highest mathematical honors at the University of Oxford. He undertook to point out that (what he has been pleased to term) my highly ingenious reasoning in the pamphlet, rests on a fallacy. Our readers will judge, after perusing the correspondence, how far he has succeeded in accomplishing the task.

My theory is, that in every circle, *the circumference is exactly equal to three and one-eighth times its diameter; and the area exactly equal to three and one-eighth times the area of a square described on its radius*; and I have demonstrated these facts by a variety of diagrams, and could have adduced many more, had I thought it necessary.

I submit, that my correspondent has entirely failed to subvert this theory, and for the best of all reasons: "*It is one of the great truths of nature, which can admit of no doubt, and which it is not in the power of any man living to subvert.*" The latter sentence I quote from the letter of another correspondent, only recently received, who, in all his former communications had, with great fairness and candor, conscientiously opposed me.

Previously to the twenty-ninth meeting of the British Association, at Aberdeen, in 1859, my enquiries had been confined to an examination of the relations existing between circles and squares, but even this limited condition of the enquiry had thoroughly satisfied me, as to the true relation between the diameter and circumference of a circle; and I resolved to bring the subject under the notice of the Association, of which I have been a member almost from the earliest period of its existence.

A short time before the meeting of the Association, I addressed a letter to the Honorary Secretary, informing him that it was my intention to attend the meeting, and that I purposed reading a Paper "On the true Circumference and Area of a Circle." I received a reply from him, to say my Paper was placed on the books, and requesting me to inform him, on what day I should be prepared to read it. To this I replied, that I should be prepared to read it on the first day of the meeting of the Sections, if necessary, or on any subsequent day if more convenient to the Association.

I was told by several of my acquaintance that the Association would never give me a hearing, and that if I wished to spare myself considerable annoyance I had better stay at home. I felt that the subject was of too much importance to the interests of science, to justify me in permitting it to be stifled, without any effort on my part to prevent it, and I resolved, at all risks, to attend the meeting of the Association.

I had the pleasure to hear His Royal Highness the Prince Consort, the President of the Association for that year, deliver his opening address, and I shall ever remember the gratification I felt, on hearing His Royal Highness make the following remarks:—"*Remembering that this Association is a popular Association, not a secret confraternity of men, jealously guarding the mysteries of their profession, but inviting the uninitiated, the public at large, to join them, having as one of its objects, to break down those imaginary and hurtful barriers, which exist between men of science and so-called men of practice.*" Surely, thought I at the moment of hearing these words, the friends who advised me to stay at home must have been mistaken. Little did I suppose, that before the end of the meeting, I should discover the practice of the Association to be so widely at variance with the theory of its constitution, as set forth in such flattering terms in the opening address of its Royal President.

The following morning I presented my Paper to the Committee of the Mathematical Section, and was told without any hesitation, I could not be permitted to read it; that the subject was prohibited from being introduced into the deliberations of the Association. For a short time, and with considerable earnestness, I endeavored to reason with the Committee as to the propriety of such a determination; but the Association had evidently arrived at the same conclusion as the late Baden Powell, viz., That "*no amount of attestation from innumerable and honest witnesses, would ever convince any one versed in mathematical science, that a person had squared the circle;*" and the attestation of so obscure and unknown an individual as the writer, in all probability produced in the minds of the Committee a feeling of the most profound contempt. Be this as it may, I reasoned in vain.

I then changed my tactics, and prepared a short Paper, entitling it, "*On the Relations of a Circle inscribed in a Square,*" and I again presented myself before the Committee, but with no better result. Subsequently, however, through the intervention of an individual member of the Committee, (J. Pope Hennessy, Esq., M.P. for King's County, Ireland, and to whom I take this opportunity of tendering my grateful acknowledgments,) my second Paper was inserted for reading in the programme of the following day, but was placed the last on a list of thirty Papers. All these Papers could not be disposed of that day, and many of them had to stand over till the next sitting of the Section, mine being among the number. What was my astonishment on finding it asserted in the programme of the following morning, that my Paper had been read the preceding day. This appeared to me to be positive evidence of a determination to burke the subject, by any means, however dishonest, and I at once resolved upon my course. I took my seat in the Section, and waited until Sir William Rowan Hamilton, the Astronomer Royal of Ireland, took the chair for the day, in the absence of Lord Rosse. I then rose, and made my complaint, demanded to read my Paper, and gave the Section to understand, that I was not the man that would permit even the British Association to trifle with me. It was not an every-day scene in the Sections of the British Association, and Sir W. R. Hamilton will not have forgotten the circumstance. He permitted me to read the Paper. (See Appendix B.) Though short, it introduces to notice some of the fundamental truths of this important discovery.

I left a written copy of the Paper with the Secretary to the Section, and subsequently forwarded a printed one for insertion in the Transactions of the Association. If the reader will refer to the Report of the Association for 1859, he will find it recorded in the Transactions of the Mathematical Section, page 10, that J. Smith read a Paper "*On the relations of a Circle inscribed in a square.*" I should like to know how much wiser any reader of the Transactions of the Association would be, for this wonderful piece of information. Could this learned confraternity have devised any better method, of jealously guarding the mysteries of their profession? or, could they have afforded better evidence of the difference between the practice of the Association and the theory of its constitution, as enunciated by its Royal President?

During the course of the meeting, I accidentally met Mr. Airy, the Astronomer Royal of England, in the quadrangle of the College. I was not personally acquainted with him, but assuming that for the time being the members of the Association met on a footing of equality, I ventured to address him, respectfully asking him if he could afford me a few minutes' conversation on an important mathematical subject, stating that I believed I had made the discovery of the true Circumference and Area of a Circle. "*It would be a waste of time, Sir, to listen to anything you could have to say on such a subject,*" was his reply, and he attempted to change the conversation, by putting a frivolous question to me on an entirely different matter. I at once observed that I was considered guilty of an intrusion, begged Mr. Airy's pardon for it, and bid him "Good Morning."

On my return home, I commenced an enquiry into the relations of a Circle with a variety of other Geometrical figures, and very soon made some most important discoveries. I then felt justified in addressing the following Letter (with some slight verbal alterations) to Sir William Rowan Hamilton:

Barkeley House, Seaforth, near Liverpool,
14th February, 1860.

Sir William,

This day twelve months, I published a small Pamphlet on "*The true Circumference and Area of a Circle,*" and on the last day of sitting of the Mathematical Section of "The British Association for the Advancement of Science," at its meeting at Aberdeen, you presided in the absence of Lord Rosse, and you may remember, that after very considerable opposition and difficulty, I did succeed in reading a Paper, entitled, "*On the relations of a Circle inscribed in a Square.*" I subsequently sent you, and at the same time I sent to many others whom I thought likely to take some interest in the subject, a copy of the Paper I had read. (Duplicate enclosed.)

I have not received a single private communication, in reference to it, and I am not aware it has ever been publicly noticed, except in one instance. The Editor of the *Athenaeum*, did condescend to inform his readers, that I had read such a Paper at the meeting of the Association, in a paragraph of three lines, not more remarkable for their falsehood, than their absurdity.[1]

I spoke to your Royal Brother of England when in Aberdeen, who dismissed me most unceremoniously, giving me very plainly to understand, *that in his opinion, it would be a waste of time to listen to anything I could have to say on such a subject.* It is just possible, you may hold a similar opinion, but as it is my intention to publish on this interesting and important question, before doing so, I feel that it is due to you, having read my Paper in the Section at which you presided, again to direct your attention to the subject, and give you an opportunity of communicating with me, if you should think proper to do so, either admitting or denying the truth of the hypothesis I assume.

I have since the period referred to, examined the properties of a Circle, not only in its relation to a circumscribed and inscribed square, but also in its relation to other commensurable geometrical figures, and I send you herewith one

---

[1] "British Association, Section A, 'On the relations of a circle inscribed in a square.' The author enunciated a few well-known relations in imperfect decimal expression, derived from the approximate numerical expression for the circumference of a circle."—*The Athenaeum*, 8th October, 1859.

instance, as an illustration of the truth of my theory. It is only one out of many I shall produce, all equally demonstrative.

I need not point out to you as a piece of information, that if the two sides of a right-angled triangle adjacent to the right angle, be in the ratio of 3 to 4, the triangle is commensurable, and the value of the third side, can be arithmetically given, either in whole numbers, or in perfect decimal expression.

But at this point, I beg to draw your attention to the fact, that this commensurable right-angled triangle, is a link in the chain of commensurable geometrical figures, connecting one with another in a most remarkable manner, so much so, that their connection may be reduced to a system, at once beautiful, harmonious, and thoroughly self-consistent.

In the following diagram, (see figure A) let ABC be a commensurable right-angled triangle, of which the side BC is to the side AB, in the ratio of 3 to 4; and D, a circle, of which the diameter is equal to the perpendicular of the triangle.

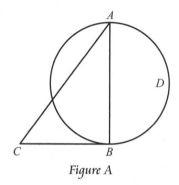

Figure A

In the first place, let the diameter of the circle, and perpendicular of the triangle, be 8.

On the orthodox hypothesis, $8 \times 3.1416 = 25.1328$, will be the approximative value of the circumference of the circle.

On my hypothesis, which, for the sake of distinction, I shall call the heterodox hypothesis, $8 \times 3.125 = 25$, will be the exact circumference of the circle.

The perpendicular of the triangle is 8. Therefore, $(3/4)(8) = 6$, will be the base of the triangle. $(5/4)(8)$, or $\sqrt{6^2 + 8^2}$, $= 10$, will be the hypotheneuse of the triangle. And $6 + 8 + 10 = 24$, will be the value of the perimeter of the triangle.

Then, on the former hypothesis, the ratio between the circumference of the circle, and perimeter of the triangle, will be, as 25.1328 to 24. And on the latter hypothesis, the ratio between the two, will be, as 25 to 24. And these ratios hold good, no matter how peculiar may be the decimal selected as the diameter of the circle and perpendicular of the triangle.

For example, let the diameter of the circle, and perpendicular of the triangle, be 7.7.

On the orthodox hypothesis, $7.7 \times 3.1416 = 24.19032$, will be the approximative value of the circumference of the circle.

On the heterodox hypothesis, $7.7 \times 3.125 = 24.0625$, will be the exact value of the circumference of the circle.

The perpendicular of the triangle is 7.7. Therefore, $(3/4)(7.7) = 5.775$ will be the base of the triangle; $(5/4)(7.7)$, or $\sqrt{(5.775^2 + 7.7^2)}$, $= 9.625$, will be the hypotheneuse of the triangle; and $5.775 + 7.7 + 9.625 = 23$ will be the value of the perimeter of the triangle; and, as $25.1328 : 24 :: 24.19032 : 23.1$; or, as $25 : 24 :: 24.0625 : 23.1$; and so far one hypothesis would appear to be just as good as the other.

But, reverse the operation, and let the circumference of the circle be the given quantity, say 60. Let it be required to find the values of the diameter of the circle, and the perimeter of the triangle.

On the heterodox hypothesis, $60 \div 3.125 = 19.2$, will be the exact value of the diameter of the circle, and perpendicular of the triangle; $(3/4)(19.2) = 14.4$, will be the base of the triangle; and $(5/4)(19.2)$, or, $\sqrt{(14.4^2 + 19.2^2)}$, $= 24$, will be the hypotheneuse of the triangle; and, $19.2 + 14.4 + 24 = 57.6$, will be the value of the perimeter of the triangle; and, as $25 : 24 :: 60 : 57.6$ exactly; and on this hypothesis the ratio holds good, whether the circumference of the circle be the given quantity, to find the diameter of the circle and the perimeter of the triangle; or, the perpendicular of the triangle be the given quantity, to find the circumference of the circle.

On the orthodox hypothesis, $60 \div 3.1416 = 19.0985$ &c. will be the approximative value of the diameter of the circle, and perpendicular of the triangle; $(3/4)(19.0985) = 14.323875$ will be the base of the triangle; and $(5/4)(19.0985) = 23.873125$, will be the hypotheneuse of the triangle; and, $19.0985 + 14.323875 + 23.873125 = 57.2955$ will be the value of the perimeter of the triangle. But, these figures are less than those required by the orthodox ratio. For, as $25.1328 : 24 :: 60 : 57.2956$, and it appears to me to be about as absurd to attempt to maintain the orthodox hypothesis, as it would be to maintain that $6 \times 8 = 48$, but that 8×6, is only equal to 47 and a fraction.

I have endeavored to bring this subject under the notice of several gentlemen, who pride themselves on their high mathematical attainments, and judging from the reception I have met with at their hands, I am disposed to think that the philosophical world is not much wiser now, than it was in the days of Galileo, but,

"Magna est veritas, et prevalebit,"

and it is not in the power of ten thousand men of the highest genius, by, their united efforts to annul this glorious fact.

Political, religious, and social freedom, have, however, made rapid strides even in our own day, in this privileged and happy nation, and it is now the right of a Briton, to be able to freely express an opinion, without incurring the risk of torture, or the danger of a prison; and in virtue of this privilege, I venture to address you thus freely, as a "great authority" on this subject, conceiving that the importance of it fully justifies me in doing so.

In conclusion, I may remark that my position in life is happily one of the most perfect independence. I have gone into this enquiry from a pure love of science, and a disinterested desire to promote it, and this course I shall continue to pursue. To myself personally, therefore, it is a matter of little consequence what course you may be pleased to adopt. I shall be glad however to find, that you do not consider the subject unworthy your attention, and in this respect form an honorable exception, among those of your professional brethren, with whom I have come into contact.

I remain, SIR WILLIAM,

Yours very respectfully,

JAMES SMITH

Sir Wm. R. Hamilton, L.L.D., &c.,
    Observatory, near Dublin.

P.S.—Having referred to an incident in connection with Mr. Airy, the Astronomer Royal of England, it might be said, if he were not made aware of it, that I was making a false charge, without having given him the opportunity of refuting it. To prevent this, I shall write him and enclose a copy of this letter, and if he should have changed his opinion, it will also give him the opportunity of admitting it.

I received from Sir W. R. Hamilton, by return of post, the following very courteous reply:

Observatory, near Dublin,
February 15th, 1860.

Sir,

I have received your letter, with its printed enclosure; another copy of this latter, (namely of the printed Paper,) had indeed reached me some months ago; but I did not understand that you required me to acknowledge it: nor would it have been a pleasant task to inform an ingenious gentleman, without necessity, of my entire disagreement from his views.

But since, while reminding me of my having had the honor to preside in the absence of Lord Rosse when you read your paper in Section A of the British Association, at Aberdeen, last year, you are pleased "again to direct my attention to the subject, and to give me an opportunity of communicating with you, if I should think proper to do so, admitting or denying the truth of the hypothesis you assume," in extracting which passage from your letter, I merely make such substitutions as that of me for you, &c. I suppose that you might be more displeased by my remaining silent, than by the distinct expression of my dissent.

I understand you to maintain, that if the diameter of a circle be represented by the number 8, the circumference of the same circle will then be represented, without any error, by the number 25. Now, Sir, I do not expect you to attach the slightest weight to any opinion of *my own*, on this or any other subject. But it will much surprise me, if you shall not find Mathematicians *unanimous* in their rejection of that result. That 8 circumferences of a circle *exceed* 25 diameters, is (I conceive) a theorem as completely certain and established in mathematics, as that the three angles of a plane triangle are together equal to two right angles.

When you next publish, if you shall think it needful to mention the fact of my having been in the chair of the Section, I hope that you will be so good as to state, at the same time, that my opinion, or rather conviction upon the subject, is entirely opposed to your own.

I have the honor to be, Sir,

Your obedient servant,

WM. ROWAN HAMILTON.

James Smith, Esq.,

Barkeley House, Seaforth, near Liverpool.

I addressed the following note to Mr. Airy, enclosing copy of my letter to Sir W. R. Hamilton:—

Barkeley House, Seaforth
Liverpool, 14th Feby., 1860.

Sir,

I herewith enclose you a copy of a letter I have this day addressed to Sir Wm. R. Hamilton.

As I have referred in it to a circumstance in which you are personally concerned, I consider it a matter of courtesy to make you acquainted with all I have said respecting you.

I am, Sir,
Yours very respectfully,
JAMES SMITH.

G. B. Airy, Esq., F. R. S., &c.,
Royal Observatory, Greenwich.

The following was Mr. Airy's reply:—

Royal Observatory, Greenwich,
15th February, 1860.

Sir,

I have this day received your letter of the 14th Inst., enclosing copy of one addressed to Sir W. R. Hamilton.

As regards Sir W. R. Hamilton, I have no remark to offer. As regards myself, you will doubtless remark, that every person has a right to publish his own views, by any inoffensive method which he may think best, but that this gives him no command, as by right, of the most valuable possession of other persons, namely, their time.

I am, Sir
Your obedient servant,
G. B. AIRY.

James Smith, Esq.

Such are some of the obstacles I have had to contend against, in announcing the discovery of a glorious scientific truth; and such, for some time to come, may probably be the fate of others, whose views, however true and important, may happen to run in antagonism to the prejudices of the scientific world. But truth, spring from whence it may, will ultimately triumph in spite of all opposition. I am not a young man, and may not live to see it, but the day will arrive, when truth by her own inherent powers shall rise above the horizon, and shining forth in all her brightness, shall dispel the darkness which now reigns on this and kindred subjects; and shall, on the one hand, "*reduce*" to their natural and proper level, that "learned fraternity" who at present contrive, like the astrologers of old, to make a mystery of their craft, and to jealously guard it; and shall, on the other hand, "induce" thousands of persons of both sexes to embrace the study of the sublime and glorious works of our Creator, as manifested in the wonderful evolutions of the heavenly bodies, who at present are intimidated into the belief, that the study of Astronomy is a something utterly beyond the capacity of the average intellect of mankind.

I may, on a future occasion, direct public attention to the importance of this discovery, in its practical application to Astronomical, Nautical, and Mechanical Science.

JAMES SMITH.

Barkeley House, Seaforth,
    Liverpool, 1st April, 1861.

Airy had it right. What circle-squarers and other cranks do, if they can, is steal time. They try to snare mathematicians into correspondence. They send out 1500 copies of their pamphlets: if each recipient devoted only a minute to it, there is half a week's worth of useful work, gone forever.

Smith deserves a little sympathy, but not much. It was not nice of the British Association to try and keep him from delivering his paper. Airy was, in his English way, rude. Cranks can and should be treated gently. To do otherwise can inflame their paranoia.

20

Legislating π

Everyone knows that Indiana (or perhaps it was Illinois, or Iowa—one of those states around there) once passed a law specifying the value of π. As with many things that everyone knows, this is false. Though the legend will probably never go away, here is one more attempt to kill it off. There follows the story of π's appearance in the Indiana legislature.

Indiana's Squared Circle

Arthur E. Hallerberg (1977)

In 1897 the Indiana state legislature considered House Bill No. 246—"a bill for an act introducing a new mathematical truth"—which proclaimed a new way of squaring the circle with an accompanying declaration of a new value of π. This amazing story was first fully documented in 1935 by W. E. Edington in a paper [2] that gives a valid account of the legislative action. Upon its introduction in the House, the bill was initially sent to the Committee on Swamp Lands; in the Senate it first went to the Temperance Committee. The Education Committee was called into the picture, and, upon the advice of the State Superintendent of Public Instruction, supported the bill. While the House passed the bill unanimously without even waiting for its third reading, as required by the state constitution, the Senate finally voted to postpone action on the bill indefinitely. Fortunately for the people of Indiana, the "indefinitely" still continues!

The author of the bill was E. J. Goodwin, M.D., of Solitude, Posey County, Indiana. It was presented to the House by representative Taylor I. Record, who was to admit that he really knew nothing about the merits of the bill, but had introduced it at the request of the country doctor who had been practicing in Posey County for nearly 20 years. (Goodwin at that time was at least 68 years old—possibly 72.)

The bill (see facing page) appears to propose two different values for π, first the value 4, and then 3.2. The 3.2 value is readily obtained by inverting the "five-fourths to four" ratio which itself is irresistible if one accepts two of Goodwin's assumptions. He speaks of them as "revealed"; although they are mathematically false, as approximations they are not too bad. Goodwin states that the ratio of chord to arc of ninety degrees is as seven to eight (Figure 1) and the ratio of diagonal to side of a square is as ten to seven. Hence, for a quadrant of 8 we have a circumference of 32 and a diameter of 10; the ratio 10/32 yields a ratio of 5/4 to 4.

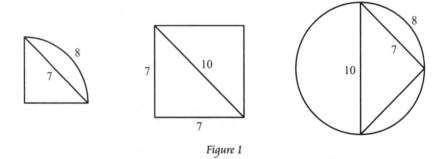

Figure 1

The value of 4 is deduced from the proportion stated in Section 1. Since an "equilateral rectangle" is simply a square, this proportion is Area of circle : Quadrant^2 = Area of Square : Side^2. Since the right side reduces immediately to 1, we appear to have $\pi r^2 = (2\pi r/4)^2$ or $\pi = 4$.

Some clarification can be obtained from Goodwin's contribution to the July 1894 issue of the *American Mathematical Monthly*. (This issue was only number 7 of volume 1 of an infant journal published independently by its two editors, B. F. Finkel and J. M. Colaw. The Mathematical Association of America, which today publishes the *Monthly* as one of its official journals, did not come into existence until 1915.) In the "Queries and Information" section, Goodwin's "contribution to science" reads as follows:

House Bill No. 246, Indiana, 1897

A bill for an act introducing a new mathematical truth and offered as a contribution to education to be used only by the State of Indiana free of cost by paying any royalties whatever on the same, provided it is accepted and adopted by the official action of the legislature of 1897.

Section 1. Be it enacted by the General Assembly of the State of Indiana: It has been found that a circular area is to the square on a line equal to the quadrant of the circumference, as the area of an equilateral rectangle is to the square on one side. The diameter employed as the linear unit according to the present rule in computing the circle's area is entirely wrong, as it represents the circle's area one and one-fifth times the area of a square whose perimeter is equal to the circumference of the circle. This is because one-fifth of the diameter fails to be represented four times in the circle's circumference. For example: if we multiply the perimeter of a square by one-fourth of any line one-fifth greater than one side, we can in like manner make the square's area to appear one fifth greater than the fact, as is done by taking the diameter for the linear unit instead of the quadrant of the circle's circumference.

Section 2. It is impossible to compute the area of a circle on the diameter as the linear unit without trespassing upon the area outside of the circle to the extent of including one-fifth more area than is contained within the circle's circumference, because the square on the diameter produces the side of a square which equals nine when the arc of ninety degrees equals eight. By taking the quadrant of the circle's circumference for the linear unit, we fulfill the requirements of both quadrature and rectification of the circle's circumference. Furthermore, it has revealed the ratio of the chord and arc of ninety degrees, which is as seven to eight, and also the ratio of the diagonal and 'one side of a square which is as ten to seven, disclosing the fourth important fact, that the ratio of the diameter and circumference is as five-fourths to four; and because of these facts and the further fact that the rule in present use fails to work both ways, mathematically, it should be discarded as wholly wanting and misleading in its practical applications.

Section 3. In further proof of the value of the author's proposed contribution to education, and offered as a gift to the State of Indiana, is the fact of his solutions of the trisection of the angle, duplication of the cube and quadrature of the circle having been already accepted as contributions to science by the American Mathematical Monthly, the leading exponent of mathematical thought in this country. And be it remembered that these noted problems had been long since given up by scientific bodies as unsolvable mysteries and above man's ability to comprehend.

QUADRATURE OF THE CIRCLE

By EDWARD J. GOODWIN, Solitude, Indiana

Published by the request of the author.

(Copyrighted by the author, 1889. All rights reserved.)

A circular area is equal to the square on a line equal to the quadrant of the circumference; and the area of a square is equal to the area of the circle whose circumference is equal to the perimeter of the square.

To quadrate the circle is to find the side of a square whose perimeter equals that of the given circle; rectification of the circle requires to find a right line equal to the circumference of the given circle. The square of a line equal to the arc of 90° fulfills both of the said requirements.

Goodwin's original (copyrighted!) premise was that a given "perimeter" defines the same area whether related in form to a circle or to a square. This is stated so clearly here and in a number of his other writings, prior to 1897, that one wonders why it was not so given in the bill itself. Perhaps the "protection of the copyright" was thought to be involved—but the proportion initially given in the bill actually says the same thing. The contribution states incorrectly the quadrature problem—which is to find the side of a square whose area (not perimeter) equals that of the given circle. At no place in any of Goodwin's writings is there any reference to the historical context of the classical quadrature problem of finding such a side by a geometrical construction using only compass and straightedge.

Returning to the bill itself, we find a phrase repeated a number of times which at first seems to make little sense: "The diameter employed as the linear unit according to the present rule in computing the circle's area is entirely wrong ...". For many mathematical neophytes, an area is something found by multiplying two lengths together. To find the area of a circle, Goodwin apparently assumed that one of the lengths should be the quadrant of the circle. Correct mathematics says: $A = \pi r^2 = (2\pi r/2) \cdot r = (C/2) \cdot (d/2) = (C/4)d$. Hence the true area is indeed $(C/4)d$, and the diameter is "the linear unit". But for Dr. Goodwin the area was $(C/4) \cdot (C/4)$.

In Section 2 Goodwin points out another bit of evidence for the "incorrectness" of the rule in present use: "... because the square on the diameter produces the side of a square which equals nine when the arc of ninety degrees equals eight". In the *Monthly* note, as well as in a number of other references, Goodwin states his case in much more dramatic terms. "It is not mathematically consistent that it should take the side of a square whose perimeter equals that of a greater circle to measure the space contained within the limits of a less circle. Were this true, it would require a piece of tire iron 18 feet to bind a wagon wheel 16 feet in circumference".

Goodwin's reasoning on the 18 foot rim for a 16 foot wagon wheel is a bit tricky, but it goes like this: Consider a circle of circumference 16. Goodwin assigns it first the diameter of 5.0929 (which is correct, based on $\pi = 3.1416$). This leads to an area for the circle, "by the rule in use", of 20.3716 (i.e., 1/4 of 16 times the diameter). Hence, if a square has this same area, the square's side would be $\sqrt{20.3716}$, which is 4.51. Now the circle with quadrant 4.51 has a circumference of 18, and so "it would require a piece of tire iron 18 feet to bind a wagon wheel 16 feet in circumference". For Goodwin it is the "rule in use" which has led, incorrectly, to this false conclusion—since his rule is correct!

Goodwin's ultimate prescription for finding the area of a circle was to multiply the diameter by 3.2, then take a fourth of this (the quadrant), and finally to square this. To find the volume of a sphere (given elsewhere, but not in the bill), find the quadrant of the circle (sphere) and then cube it. This is good reasoning by analogy, if not good mathematics.

One might well wonder how results based on such reasoning could appear in a journal devoted to advanced mathematics. As noted before, the *Monthly* was in its first year of publication and the editors/publishers were both looking for material and promoting interest in the publication. On the other hand, the notation "by request of the author" is very seldom found with other contributions in these early years. Albert A. Bennett, in speaking of the *Monthly* before it was taken over by the Association, refers to "the largely rustic quality of early issues" [1]. Insight into the policies of the editors can be obtained by noting a large running debate on the validity of non-Euclidean geometry in early issues of the journal. In November 1895, Editor Finkel wrote: "Dr. (John N.) Lyle may be regarded as the greatest Anti-Non-Euclidean Geometer in America, and he has furnished many papers for publication in the *Monthly*. These we shall publish as our space permits. Truth has nothing to fear at the hands of any one, and if the Non-Euclidean doctrine is true, Dr. Lyle's papers will only aid in establishing it. … The editors of the *Monthly* belong to the Non-Euclidean school of thought".

No rebuttal or criticism of Goodwin's contributions ever appeared in the *Monthly*. But in December 1895, David Eugene Smith contributed an article, "Historical Survey of the Attempts at the Computation and Construction of π", which was actually a translation from Felix Klein's *Vorträge über ausgewählte Fragen der Elementargeometrie*; this of course included reference to Lindemann's proof of the transcendence of π.

When interviewed by a reporter for the Indianapolis *Sun* immediately after his proposal had been adopted unanimously by the House, Goodwin confessed that he had never devoted much time to mathematics, and only since 1888 had he thought about the circumference of the circle. The reporter noted that in Goodwin's room in Indianapolis were "his pamphlets on the relativity of relations, etc., etc., solving problems of life and the universe that have made the sages of all times scratch their heads until they became bald. 'If I were to say that the discoveries are revelations to me, they wouldn't believe it. This is an age of unbelief. Do you know it?'" [7].

Goodwin's original interests were in the combination of science-philosophy-religion that was not uncommon in those days. He published at least three versions of a monograph entitled *A New Physical Truth* in 1884, 1885, and 1892. The last version bore the title *Universal Inequality is the Law of All Creation*. Its major thesis was his One Law of the Universe: "All change depends on an inequality in the adjustment of forces whereby particles and aggregates compress to and repel from centres while acting in lines least resisting" [4]. In the third version (but not in the earlier two) we find several pages devoted to the quadrature of the circle. The mathematical material is essentially a preliminary version of that presented in the *Monthly* and in the bill itself.

Towards the end of the booklet we find the following autobiographical comment:

> During the first week in March, 1888, the author was supernaturally taught the exact measure of the circle, just as he had been taught three years before, the "Scheme of Universal Creation". These revelations were due in fulfillment of Scriptural statements and promises. Mathematicians second to none in this country, frankly admit that no authority in the science of numbers can tell how the ratio was discovered .... To assert that my experience differs from that of any other man, is, to say the least, a declaration of no common import.... All knowledge is revealed directly or indirectly, and the truths hereby presented are direct revelations and are due in confirmation of scriptural promises.

In this regard, it should be stated that Goodwin's contribution is by no means unique. One can easily document stories of half-a-dozen persons in England, Germany, and the United States who similarly received "divine revelations" of the true value of the circumference-diameter ratio—none of which was 3.14159....

While Goodwin at first felt his revelations concerning the circle were primarily to substantiate his "physical truth" concerning Universal Inequality as the Law of all Creation, as time went on he devoted more and more attention to the quadrature problem. His solution was copyrighted in England as well as in the United States. He corresponded with professors at the National Observatory in Washington, D.C., and was convinced that he had convinced the astronomers there of the truth of his results. He called upon the officers of the Smithsonian Institution to award prizes of $10,000 each (which he would supply) to any scientist who proved able to find an exception to the statement of his new physical truth, or to any mathematician who proved able to ascertain the means whereby the new ratio was found. He attempted to give a lecture-demonstration of his work at the 1893 World's Fair in Chicago.

Knowledge of these activities, all of which occurred prior to the legislative action of 1897, gives some insight into the dilemma confronting these legislators. It seems evident that nearly all of the legislators were completely unknowledgeable about the underlying mathematics and the import of the bill. We now know that almost immediately an Indianapolis daily paper exposed the absurdity of the bill by giving a brief resume of the history of π, including reference to the Rhind Papyrus, Archimedes, Huygens, Lambert, Lindemann, and Weierstrass. The only drawback was that the newspaper was *Der Tägliche Telegraph* [8] which was published entirely in German! On the other hand, the recognition that Goodwin had obtained in prior years seemed authoritative to some, particularly to members of the various committees with which he apparently met. Indicative of this feeling is an editorial which appeared in the Indianapolis *Journal* [6] after the bill was postponed, which illustrates how easily misinterpretations of earlier misunderstandings can become magnified.

> Some newspapers have been airing their supposed wit over a bill introduced in the Legislature to recognize a new mathematical discovery or solution of the problem of squaring the circle, made by Dr. Goodwin, of Posey County. It may not be the function of a Legislature to endorse such discoveries, but the average editor will not gain much by trying to make fun of a discovery that has been endorsed by the American Mathematical Journal; approved by the professors of the National Astronomical Observatory of Washington, including Professor Hall, who discovered the moons of Mars; declared absolutely perfect by professors at Ann Arbor and Johns Hopkins Universities; and copyrighted as original in seven countries of Europe. The average editor is hardly well enough versed in high mathematics to attempt to down such an array of authorities as that. Dr. Goodwin's discovery is as genuine as that of Newton or Galileo, and it will endure, whether the Legislature endorses it or not.

There were a number of factors which combined to cause the indefinite postponement of the bill. Editorial ridicule in various newspapers, out-of-state as well as local, undoubtedly played a role. The opportune appearance and "coaching" of Purdue mathematician C. A. Waldo [9] must have influenced a few of the legislators; however, his claim that "it was probably (this) alone" that prevented its adoption simply is not supported in any other account.

The final report on the bill given by the Indianapolis *Journal* (February 13, 1897) is probably the most valid overall assessment of the Senate's action:

> Although the bill was not acted on favorably, no one who spoke against it intimated that there was anything wrong with the theories it advances. All of the senators who spoke on the bill admitted that they were ignorant of the merits of the proposition. It was simply regarded as not being a subject for legislation.

We conclude this look at Goodwin's life and thought by quoting his obituary from the New Harmony, Indiana *Times*, June 27, 1902. It summarizes his entire life rather well:

End of a Man Who Wanted to Benefit the World

> Dr. E. J. Goodwin died at his home in Springfield Sunday, aged 77 years. He had been in feeble health for some time, and death came at the end of a long season of illness. Dr. Goodwin was no ordinary man, and those meeting him never failed to be inspired by this fact. He was of distinguished appearance and came from Virginia where he received an excellent education. He has devoted the last years of his life in an endeavor to have the government recognize and include in its schools at West Point and Annapolis his method of squaring the circle. He wrote a book on his system and it was commented on largely and received many favorable notices from professors of mathematics.
>
> He felt that he had a great invention and wished the world to have the benefit of it. In years to come Dr. Goodwin's plan for measuring the heavens may receive the approbation which was untiringly sought by its originator.
>
> As years went on and he saw the child of his genius still unreceived by the scientific world, he became broken with disappointment, although he never lost hope and trusted that before his end came he would see the world awakened to the greatness of his plan and taste for a moment the sweetness of success. He was doomed to disappointment, and in the peaceful confines of village life the tragedy of a fruitless ambition was enacted.

There is always the human element in mathematics, whether the mathematics is right or wrong!

References

[1] A. A. Bennett, Brief History of the Mathematical Association of America Before World War II, *Amer. Math. Monthly*, 74 (1967) 8 (Part II, Fiftieth Anniversary Issue).

[2] W. E. Edington, House Bill No. 246, *Proc. Indiana Acad. Sci.*, 45 (1935) 206–210.

[3] E. J. Goodwin, Quadrature of the circle, *Amer. Math. Monthly*, 1 (1894) 246–247.

[4] ———, Universal Inequality is the Law of All Creation, Solitude, Indiana (1892) 61–62.

[5] A. E. Hallerberg, House Bill No. 246 Revisited, *Proc. Indiana Acad. Sci.*, 84 (1975) 374–399.

[6] The Indianapolis *Journal*, Indianapolis, Indiana, February 22, 1897.

[7] The Indianapolis *Sun*, Indianapolis, Indiana, February 6, 1897.

[8] *Der Tägliche Telegraph*, Indianapolis, Indiana, January 20, 1897.

[9] C. A. Waldo, What might have been, *Proc. Indiana Acad. Sci.*, 26 (1916) 445–446.

I thought that it would be a good idea to visit Solitude, Indiana, Dr. Goodwin's town, to gather local color. Though it was not likely that any of his relics would survive, one never knows what might turn up.

Alas, serendipity sometimes fails to operate properly. My first difficulty was in finding Solitude. It is on the Indiana highway map, but it did not seem to be where it should have been. The difficulty was that a sign was missing. The state of Indiana posts green and white signs at the city limits of many of its cities and towns, but the one on the road approaching Solitude from the north was missing. The one on the south side of town was there, though, so Solitude was findable.

Not much of it survives. There are a dozen or so houses and that's all. There are no businesses, no post office, and certainly no doctor's office. Solitude is one of those many towns made redundant by the automobile, and though it may have been bustling in Dr. Goodwin's day, now it is gone. None of the houses dated back to 1902, so I did not stop at any to inquire about him.

There is an excellent paper by David Singmaster, "The legal values of pi" in the *Mathematical Intelligecer* 7 (1985) #2, 69–72. Professor Singmaster reads Goodwin closely and finds many different values for pi, ranging all the way from 2.56 to 4, implicit in his confused exposition.

Arthur Hallerberg received his doctoral degree from the University of Michigan in 1957 and taught at Valparaiso University for many years. He was one of the editors of the National Council of Mathematics's thirty-first yearbook, *Historical Topics for the Mathematics Classroom*. He died in 1978, aged 60.

21
Mathematics and Music

What, exactly, is the connection between mathematics and music? It seems to be universally agreed that there is one, but its nature has not yet been made clear, though not for lack of trying. Edward Rothstein, Critic at Large for *The New York Times* and not unacquainted with mathematics, made an effort in *Emblems of Mind* (subtitle, "The Inner Life of Music and Mathematics") from which the selection that follows is taken.

Music, mathematics, and chess are where early genius is found, thus implying that they have something in common. We have Mozart composing while still a boy, Pascal proving his hexagon theorem at the age of sixteen, and Capablanca improving on his father's chess moves when he was hardly tall enough to see over the edge of the table. There are no child prodigies in, say, metallurgy. The explanation probably is that music, mathematics, and chess all involve manipulating symbols in a self-contained system. No knowledge of the world or extensive experience is needed to see, if your mind has the capacity, what should be the next move on a chessboard, the next step in an argument, or the next note in a sonata. It's reasonable that minds with the capacity to manipulate symbols in one field should be able to do it in others. Mathematicians tend to like music and quite a few musicians like to play chess. Top-ranked chess players seem to be too single-minded to bother with anything else, but, if they had the time, they might be dabblers in the other two fields.

It may be that early evidence of talent is all that mathematics and music have in common. I tend to think so, though Mr. Rothstein probably would disagree. If mathematics and music were closely related, we would expect developments in one field to parallel those in the other, but this has not happened—in fact, just the opposite has occurred. When mathematics was undergoing its romantic period, when Euler was freely writing things like $-1 = 1 + 2 + 4 + 8 + \cdots$

(letting $x = 2$ in $\frac{1}{1-x} = 1 + x + x^2 + x^3 + \cdots$), the age of Bach had just passed and Mozart and Haydn were writing their classical works. When mathematics was becoming formal and rigorous in the nineteenth century, with all those epsilons and deltas, wild men like Berlioz and Liszt were breaking the bounds of musical restraint. The 20th century was one of tremendous mathematical accomplishments, but not so, I think, in music.

Further, mathematics and music aim at different targets, mathematics at the head and music at the heart. Although mathematicians are fond of mentioning the beauty of this theorem or that, the beauty is not the same as musical beauty. We do not give ourselves over to mathematical beauty, we do not let it wash over and ravish us, it does not stir us to the core. It is there, but it is more or less incidental. Similarly for the intellectual appeal of music. Not many of us study scores, appreciating the thought and skill that has gone into them. The thrust of music lies elsewhere.

Mr. Rothstein points out some connections in his book, such as

The language of each is difficult to decipher, yet rich in information, and often notation itself leads to new discovery.

Both fields have an emphasis on mappings, proportions and relationships.

Both have a sense of style.

Here is a representative part of the book.

$-\!\!\!\sim\!\!\!\mathcal{O}\!\!\!\sim\!\!\!-$

Emblems of Mind

Edward Rothstein (1995)

Fugue: The Making of Truth

When judging a physical theory, I ask myself whether I would have made the Universe in that way had I been God.   —Albert Einstein

Though the end of our journey is in sight, the different paths we have taken are still shrouded in mist. Though the earth may occasionally be bright at our feet, we have yet to see sunrise from the top of our Mt. Snowdon, let alone the vision that awaited the poet at the end of his climb. For while we have begun to understand the nature of musical and mathematical thinking and have touched, tentatively, on the nature of musical and mathematical beauty, the most vexing questions remain. Despite occasional glimpses of other regions, we have been preoccupied with purely formal relations, the workings of math and music at the most abstract levels. We know that these constructions involve transformations, mappings, play with resemblance and difference; we know too that in these constructions there are exquisitely unexpected alliances and connections, resonances and proportions.

But these are incomplete speculations. Games of chess can also be beautiful constructions, full of internal allusions and wit, but chess is limited in a way that music and mathematics are not. Chess is merely abstract; the game does not transform the world; a brilliant strategy does not have much impact outside the universe defined by the game. But there is something more than abstract about mathematics and music.

We already know this to be so from our speculations about how the beautiful and the sublime relate to our experience, and how some perception of beauty is related to our understanding of the truth. But our understanding is still imprecise and may be doomed to remain so. Why did the Pythagoreans believe music and mathematics to have connections both inward and outward —affecting the currents of our souls and the structure of the universe itself? Why are mathematics and music often associated with religious ritual—giving it shape and order while attempting to invoke forces seemingly beyond reason? What gives mathematics its continued power in explaining the world in unexpected ways? And what gives music its power to inspire fear and dread and ecstasy in its listeners, leaving not even the most powerful institutions untouched by its influence? These are not questions about the internal workings of mathematics and music but about the ways they map into the world; these are questions about their meaning and truth.

The problem in mathematics can be posed differently. We know that mathematics gives us tools for organizing phenomena within the world. The puzzle is why mathematics at its most abstract turns out to have such power as well. As we have seen, systems of non-Euclidean geometry were abstract mathematical constructions before it was discovered that they could also describe the physical world. For instance, knot theory—a totally abstract exploration of knots—turned out to be an effective tool in a recent theory of the cosmos, known as string theory. Why should this be? What is the connection between brainspun speculation and brute physical fact?

One physicist, Eugene Wigner, in an essay titled "The Unreasonable Effectiveness of Mathematics," wrote, "The enormous usefulness of mathematics in the natural sciences is something bordering on the mysterious. There is no rational explanation for it." Albert Einstein asked, "How can it be that mathematics, being after all a product of human thought independent of experience, is so admirably adapted to the objects of reality? Can human reason without experience discover, by pure thinking, properties of real things?"

It is difficult to answer such questions when it is unclear even what "pure thinking" might be. There are mathematical truths so powerful and pristine they bear little resemblance to everyday truth; there are other mathematical truths that seem manifestly inadequate for any use. A syllogism may be true— "If Socrates is a man and all men are mortal, then Socrates is mortal"—but it can also be useless; it is just as true, for example, to argue: "If Socrates is a bird and all birds are immortal, then Socrates is immortal." But this is a true syllogism in which every term is false.

We cannot even take simple mathematical argument for granted in its relation to our world. Ludwig Wittgenstein—in later life a radical skeptic about logic and mathematics—argued again and again that in moving from the world of our experience into the world of mathematical reasoning and back again we are making very great leaps. We might feel compelled to come to a conclusion in mathematical and logical reasoning yet feel absolutely no such compulsion in our understanding of the world. Mathematics has a relation to experience that can be as flexible and problematic as that of language itself. It can also take on a status which no other aspect of our lives shares. How, for example, should we treat an ordinary arithmetical statement like $2 + 2 = 4$? Is it the result of an abstraction from experience, as John Stuart Mill would have asserted? If it is a hypothesis deriving from experience with real objects, then it should be subject to verification and proof from experience. But, in fact, if we saw an instance in which $2 + 2$ were not 4, we would discount our counting, not the assertion.

So confusing is the nature of mathematical truth that there are statements in mathematics that seem true but have been neither proven nor disproven. One famous example is the Goldbach conjecture, which asserts that every even number is the sum of two primes. For example, $4 = 2 + 2$; $6 = 3 + 3$; $8 = 5 + 3$. Such sums have been found for every even number up to 100 million. But nobody has established that the assertion, so simply stated, is actually true for all even numbers. Twentieth-century logic has gone further; it has shown that there are statements in mathematics which are indeed true but can never be proven.

So the minute we begin contemplating the notion of mathematical truth we run into a phantasmagorical swirl of issues philosophers have been pondering for millennia. Yet there math is, wherever we turn. We may not know what "pure thinking" is, but we do know it often comes in handy. "Whoever undertakes to set himself up as a judge in the field of Truth and Knowledge," Einstein warned, "is shipwrecked by the laughter of the gods."

Many of these problems disappear, though, if we alter our conception of those gods. There is a long tradition in mathematical thought, a tradition that goes back to its very origins in the surveying of land or the mapping of the heavens. In that tradition there is no problem at all in assessing mathematics' unreasonable effectiveness, or the nature of mathematical truth, because the world of mathematics is actually more real than the world, not less; it is primary, not secondary; it presents the essence of objects and their relations. This is not a mere metaphor. Plato maintained that individual objects only exist as imitations and reflections of a higher order of Forms. He believed the problem to be the inverse of the one we have posed; the challenge is not to explain how mathematics fits the world but how the world develops out of something resembling mathematics. Mathematics, in this view, is a divine language.

In many cultures' views of mathematics (and music), parallels exist with the world. These beliefs make Einstein's question about the connection between thinking and the world trivial: of course, thinking reflects the universe; it can do nothing else. We think the way the universe works. This was a strong tradition in Western philosophy of the last five hundred years as well. It was believed that when we discovered the properties of pure thinking, we would have a language that would automatically produce truths about real things—partly because the essence of real things was in the pure thinking. Leibniz, for one, hoped that philosophy could construct a language that would precisely reflect the universe: the only things expressible in it would also be true. The logician Gottlob Frege set out to define the laws of thought a century ago. Then Bertrand Russell sought to derive all mathematical truth out of logical truth: to prove that mathematics was founded in the laws of thought and that the laws of thought, in turn, reflected the truth. His program turned out to be unsuccessful, but the faith it represents is revealing.

These diverse efforts are rooted in the belief that the world can be described mathematically because its structure is mathematical. God, in this view, is a mathematician. In the eighteenth century—the same era that gave us the calculus and began to ask reflective questions about premises and connections between physical law and analytical thought—this belief was part of the development of modern physics. When searching for the rules that govern the revolutions of the planets, Kepler posited that each orbit literally created a pitch, a vibration in the heavens. Taken together, these pitches created a music of the spheres, a music of which his formulas were just the mathematical expression. In *Mystery of the Cosmos*, he wrote, "God himself was too kind to remain idle, and began to play the game of signatures, signifying his likeness into the world; therefore I chance to think that all nature and the graceful sky are symbolized in the art of geometry." The assertion is that God's likeness is the art of geometry, and that the art of geometry (and its reflection in music) is, in turn, the signature of God.

Galileo was still more explicit:

Philosophy [Nature] is written in that great book which ever lies before our eyes. I mean the universe, but we cannot understand it if we do not first learn the language and grasp the symbols in which it is written. The book is written in the mathematical language, and the symbols are triangles, circles and other geometrical figures without whose help it is humanly impossible to comprehend a single word of it, and without which one wanders in vain through a dark labyrinth.

This unshakable faith in the figures of geometry and formulas of mathematics guided the research itself. For centuries the circle was considered the perfect geometrical form; and since anything so perfect would be reflected in the heavens, it was thought that the motion of the stars and planets had to be circular. So strong was this belief that, though the data did not support it, the theory of planetary motion became more and more convoluted, simply to preserve the exalted status of the circle: there were circles made within circles, epicycles upon epicycles, until by the time of Copernicus the revolutions of the planets were depicted with complicated looping circular forms, dizzyingly whirling around one another, in an attempt to match the increasingly sophisticated observations of heavenly bodies.

Kepler did not discard the mathematical faith in geometrical structure; he simply tried to find a different principle of order and shape. For example, "I undertake to prove that God, in creating the universe and regulating the order of the cosmos, had in view the five regular bodies of geometry as known since the days of Pythagoras and Plato, and that he has fixed according to those dimensions, the number of heavens, their proportions, and the relations of their movements." The radii of the six then known planets, Kepler argued, should be the radii of spheres inscribed in the five regular solids: the cube, the tetrahedron, and so on.

Though the data finally did not support this elegant belief, the mathematical faith was still not discarded. In fact, Kepler spent decades of close observation, playing with numbers, determined to find in them a pattern that would be not just regular but also mathematically refined. The years of minute and laborious calculations, the obsessive attempts to fit numbers into predetermined schemes of form and beauty, finally resulted in a simple formula which spoke with the force of revelation. The planets, Kepler concluded, move in the shape of ellipses—curves that, like the circle, can be created by slicing a cone. In addition, the planets' orbits could be described according to surprisingly simple mathematical principles: the square of the period of revolution of a planet's orbit varies with the cube of its distance from the sun.

This is an extraordinary assertion; it almost seems invented with its formal precision. It is all the more remarkable that Kepler's discovery of this law came from the analysis of thousands of astronomical observations; it had nothing to do with a theory of gravity or motion. The law was not derived from some analogy (as Newton was to make) between the laws governing the motions of the heavenly bodies and the laws governing earthly ones. Kepler came upon his law because of his conviction that a mathematical relation had to govern the cosmos.

Here is Kepler's own reaction to his discovery, which is strikingly similar in many ways to Bach's reactions to his compositions: "The wisdom of the Lord is infinite; so also are His glory and His power. Ye heavens, sing His praises! Sun, moon, and planets glorify Him in your ineffable language! Celestial harmonies, all ye who comprehend His marvelous works, praise Him. And thou, my soul, praise thy Creator! It is by Him and in Him that all exists. That which we know best is comprised in Him, as well as in our vain science. To Him be praise, honor and glory throughout eternity."

There is no question that without the faith that a description of planetary motion should be simple, elegant, and fundamental—that the universe itself is ordered in such a way as to be predictable and harmonious and accessible to the powers of mind—Kepler would not have arrived at his formula. Without an abiding faith, Kepler would have had no clear way to organize the data from the thousands of astronomical sightings. His was not just a faith in the order of the universe; it was a faith in the human power to discern that order. "All pure Ideas, or archetypal patterns of harmony," Kepler wrote, "are inherently present" in the minds of those apprehending them.

Newton went further. Not only was he, like Kepler, intent on finding simple laws governing the behavior of objects but he was intent on explaining why such laws might hold. This quest involved a still greater faith. So connected were theoretical notions and philosophical ones that Newton's description of God in the *Principia* makes the deity a sort of grand Fluxion, a divine representation of the principle underlying the invention of the calculus: "He endures forever and is everywhere present; and, by existing always and everywhere he constitutes duration and space. Since every particle of space is *always*, and every indivisible moment of duration is *everywhere*, certainly the Maker and Lord of all things cannot be *never* and *nowhere*."

What amazed Eugene Wigner and other twentieth-century physicists who have wondered at the results of such thinking is that this religious faith in a mathematical order was in fact rewarded. For example, Newton asserted that gravitational force varies inversely with the square of the distance between two bodies, basing his law on the measurements of speeds of falling bodies. But there is no way he could have measured the speed of falling bodies with a sufficient level of accuracy. His measurements did not really justify his conclusion. Wigner believes that Newton's formula could only have been verified experimentally in his time within a range of 4 percent. The fact that it was asserted nonetheless, without any expectation of error, and as a law governing all bodies in the cosmos—not just objects falling to earth—illustrates Newton's essential conviction that mathematics is the language of the universe, and that this language expresses relationships simply and elegantly. When measurements were made, they were considered more fallible than the mathematical relation posited between them. Newton had no convincing notion of how gravity works, yet he was able to describe its actions as a reflection of ratio.

But more astounding than the faith is its justification by results. Wigner has written that Newton's law of gravitation has turned out to be accurate to within less than a *ten thousandth* of a percent despite the crudity of his experimentation: "The mathematical formulation of the physicist's often crude experience leads in an uncanny number of cases to an amazingly accurate description of a large class of phenomena."

Things are not all that different in today's physics, which is more dependent on mathematics than it has ever been before, and more replete with unexpected applications of mathematical theories. Indeed, there are areas in physics which are first known only through mathematical theory. The faith is not just that mathematics represents the truth but that it represents the truth so much more faithfully than reason gives us any right to expect.

We are not about to explain this uncanny efficacy of mathematics, nor do we have a theory of theories that will precisely reveal the relationship between abstraction and application. But we can begin to understand why the powers of mathematics are not as implausible as they seem. When we began our exploration into music and mathematics, we thought about the nature of comparison—noting, for example, similarities between a raven and a writing desk. We had to define the characteristics we observed in each, consider what functions they serve, and decide whether there were any significant similarities. Finding links and connections between different objects is only one of the ways we make sense out of any stream of phenomena we are exposed to, passing by us like notes in a musical score or like seemingly random collisions of sub-atomic particles. We try to find repetitions, similarities, ratios. Our search for pattern means dividing phenomena, grouping their constituent elements into smaller categories, constructing "phrases," seeing whether, when we do group them together, we recognize one element's relation to another. We must also constantly decide what is important and what is negligible, which facts have significance and which do not. We do the same thing when we try to make sense of music.

Physics must abstract from the barrage of events in the world just such essential characteristics that might repeat, follow one another, or affect one another. If a physicist gazes at the heavens, it is to note repeated patterns in the positions of the planets or define (as did primitive peoples) the outlines of con-stellations; the patterns created today may be less animistic, but the motives are the same. The very notion of natural law upon which physics is based is an as-sertion that resemblances are to be found, that there is in each event not just the particular but an example of the universal. So physics, like music, takes as its project the recognition and articulation of form in the midst of phenomena. It attempts to find analogies. It should not be surprising then that mathematics is closely aligned with physics, since mathematics is in its essence (or as close to it as we are likely to come) the study of form and structure. Mathematics takes the techniques of abstraction and observation used in physics as simply one special case of a series of methods to be applied again and again. Mathematics, almost by definition, provides the method of physics.

Theme and Variations: The Pursuit of Beauty

> *The mathematician is only complete insofar as he feels within himself the beauty of the true.*   —Johann Wolfgang von Goethe

The formal connections between music and mathematics, the ways in which each art is also a science, may now be more evident, but we must come closer to their inner lives. We must begin to see them from within, to understand not just how their processes work but how they act on their devotees and what promises they offer.

The inner life is not easily reached from outer appearance. No purely objec-tive analysis of *The Marriage of Figaro* can match the experience of Mozart's music; no description of formal relations can account for the almost transcendental qualities of the music of Bach. Robert Schumann wrote upon hearing Hector Berlioz's *Symphonie Fantastique*: "Berlioz, who studied medicine in his youth, would hardly have dissected the head of a beautiful corpse more reluctantly than I dissect his first movement. And has my dissection achieved anything useful for my readers?" An analysis presents only a single narrative about a piece that can tell many tales at once; analysis has strict limits and may even miss music's essential power. There is a vast gap between what we say and what we feel about music.

In mathematics the inadequacies of mere analysis are still more clear; as we have already seen, a three-line proof represents years of work and consideration. But at least in mathematics we can begin to get some sense of the inner life of the art through the accounts of its creators. Henri Poincaré, in a lecture about mathematical creation, told of his labors attempting to show the existence of a group of mathematical functions called Fuchsian. Here are his oft-cited words:

> Just at this time, I left Caen, where I was living, to go on a geologic excursion under the auspices of the School of Mines. The incidents of the travel made me forget my mathematical work. Having reached Coutances, we entered an omnibus to go some place or other. At the moment when I put my foot on the step, the idea came to me, without anything in my former thoughts seeming to have paved the way for it, that the transformations I had used to define the Fuchsian functions were identical with those of non-Euclidean geometry. I did not verify the idea; I should not have had time, as, upon taking my seat in the omnibus, I went on with a conversation already commenced, but I felt a perfect certainty. On my return to Caen, for conscience's sake, I verified the result at my leisure.

This tale of illumination—earlier eras might have called it revelation—is all the more famous because it is not untypical. Carl Friedrich Gauss gave a similar account of a theorem he had been unable to prove for years: "Finally, two days ago, I succeeded, not on account of my painful efforts, but by the grace of God. Like a sudden flash of lightning, the riddle happened to be solved. I myself cannot say what was the conducting thread which connected what I previously know with what made my success possible."

Jacques Hadamard, in an influential little book, *The Psychology of Invention in the Mathematical Field*, compared the sense mathematicians have of seeing the answer to a problem instantly and in its entirety with Mozart's letter about seeing his symphonies whole in his mind before setting them down on paper. The "seeing" is not literal; it is not pictorial. It is a sense instead of grasping something at once, correctly and completely.

Poincaré suggested that the vision reveals a kind of choice. The rules for such choice, he went on, "are extremely fine and delicate. It is almost impossible to state them precisely; they are felt rather than formulated." He connected insight into mathematics to "emotional sensibility":

> It may be surprising to see emotional sensibility invoked apropos of mathematical demonstrations which, it would seem, can interest only the intellect. This would be to forget the feeling of mathematical beauty, of the harmony of numbers and forms, of geometric elegance. This is a true aesthetic feeling that all real mathematicians know, and surely it belongs to emotional sensibility.

This feeling of rightness and wholeness has something to do with our ideas of beauty. And while beauty is an unquestioned aspect of music, we may learn more by following Poincaré's lead into mathematics.

We must recall, first, that there is something known as mathematical style, which we briefly explored in "Partita," the second chapter. We can speak, for instance, of the "rhythm" a proof—the ways ideas are introduced, the kinds of punctuation that are used, the imitations of generic models. There are moments of drama, surprise, even compositional styles in mathematical proof. Here is the great nineteenth-century physicist Ludwig Boltzmann on his still greater colleague J. Clerk Maxwell:

> Even as a musician can recognize his Mozart, Beethoven, or Schubert after hearing the first few bars, so can a mathematician recognize his Cauchy, Gauss, Jacobi, Helmholtz, or Kirchhoff after the first few pages. The French writers reveal themselves by their extreme formal elegance, while the English, especially Maxwell, by their dramatic sense. Who, for example, is not familiar with Maxwell's memoirs on his dynamical theory of gases? ... The variations of the velocities are, at first, developed majestically; then from one side enter the equations of state; and from the other side, the equations of motion in a central field. Ever higher soars the chaos of formulae. Suddenly, we hear, as from kettle drums, the four beats, "put $n = 5$." The evil spirit V (the relative velocity of the two molecules) vanishes; and, even as in music, a hitherto dominating figure in the bass is suddenly silenced, that which had seemed insuperable has been overcome as if by a stroke of magic. ... This is not the time to ask why this or that substitution. If you are not swept along with the development, lay aside the paper. Maxwell does not write program music with

explanatory notes. ... One result after another follows in quick succession till at last, as the unexpected climax, we arrive at the conditions for thermal equilibrium together with the expressions for the transport coefficients. The curtain then falls!

It is not necessary even to understand the details of Maxwell's argument to get a sense of its spirit from this description. It is also no accident that Boltzmann should have used the imagery of Romantic symphonic music to give an account of Maxwell's style of argumentation. Maxwell was the man, after all, who, during the same years these symphonies were being written, had composed a brilliant series of equations connecting the phenomena of electricity and magnetism with the concept of invisible "waves" in space—insights which were still resonant. There is a Romantic spirit evident behind those equations and behind Maxwell's other work; it would not be difficult to imagine accompanying sounds of kettledrums, chaos, climaxes, and a final harmonic equilibrium.

It requires intimate familiarity with the gestures of mathematical reasoning to be sensitive to such aspects of style, which are present in all mathematical work. But style is not just a casual concern of mathematicians (or mathematical physicists). What we have been calling style is actually an aspect of the emotional sensibility of mathematics, part of its inner life. Beauty is one of mathematics's goals.

That goal can even take priority over others. The mathematician Hermann Weyl wrote, "My work always tried to unite the true with the beautiful; but when I had to choose one or the other, I usually chose the beautiful." G. H. Hardy, England's leading mathematician in the first half of the twentieth century, who wrote one of the most beautiful memoirs by a mathematician, asserted more dogmatically that "there is no permanent place in the world for ugly mathematics." And Bertrand Russell, who, with Alfred North Whitehead, devoted not a little energy to the attempt to systematize all of arithmetic using the symbols and syllogisms of mathematical logic, affirmed, "Mathematics, rightly viewed, possesses not only truth, but supreme beauty—a beauty cold and austere, like that of sculpture.... The true spirit of delight, the exaltation, the sense of being more than Man, which is the touchstone of the highest excellence, is to be found in mathematics as surely as in poetry."

Anyone who has ever listened to a lecture by a physicist or mathematician whose understanding of the subject is profound as well as thorough can also provide accounts of such aesthetic appreciation. One student of the nineteenth-century mathematician Charles Hermite wrote, "Those who have had the good fortune to be students of the great mathematician cannot forget the almost religious accent of his teaching, the shudder of beauty or mystery that he sent through his audience, at some admirable discovery or before the unknown."

When I studied advanced mathematical analysis with Shizuo Kakutani and followed his extraordinarily compact arguments and elegant solutions, or heard Abraham Robinson lecture almost casually about his advances in mathematical logic, it was not the drama of the speaker's voice that caught me: in neither case was the lecture a "performance" like those familiar to students of the humanities. The lecture was an unfolding, taking place so subtly that, by the time the classes were completed, there was a sense that one had been led into regions one had not known existed. There were times when it seemed as if I had understood the lecture, as if all the details were in place, only to find out, during the following week, that the more I thought about the lectures the more intricate and subtle I realized the arguments were. In other words, I had the same sense one has when confronting an artwork whose powers seem more mysterious than those of simple propositions and whose influence can be longer lasting than any other experience of our senses. Even if we have never felt a shudder of beauty or mystery, or associated it in any way with mathematics, we can attempt to comprehend it.

Edward Rothstein is also the author of *Visions of Utopia* (2003). His current work at *The New York Times* can be accessed at

topics.nytimes.com/top/reference/timestopics/people/r/
edward_rothstein/?inline=nyt-per

22
Mathematics Books

Some people like to collect things while others do not. There seems to be no logical explanation for either sort of behavior, so there must be a collector's gene somewhere in the human makeup that, when activated, causes people to amass objects. There is no other way to explain bizarre collections such as those of telephone insulators, barbed wire, or Beanie Babies, or more common collections such as stamps, baseball cards, or Andy Warhols. The possession of *things* must soothe and expand the souls of collectors, even though they know that what they have is transitory and their collections are not likely to survive them.

Some people collect books. I have quite a few myself: I like having them, I like looking at them, sometimes even reading them, and I wish I had more. I am not alone in this feeling, as the following selection shows. (Prices have gone up since 1982, when it was published.)

On the Value of Mathematics (Books)

G. L. Alexanderson and L. F. Klosinski (1982)

Leonardo's *Codex Leicester* sold at auction at Christie's recently for $5.8 million. Experts viewed this as a bargain price and it may well be, for one has little on which to base a comparison. Few Leonardo manuscripts come up for sale. But of greater interest to mathematicians might be the question of what has happened over the years to the prices of scientific, or specifically mathematical, books and manuscripts less exalted than something in Leonardo's own hand. Have prices for this type of material risen as rapidly as those for art and antiques, as investors search for something of lasting value in times of rapid inflation? Along with increasing demand from investors one sees the parallel phenomenon of decreasing supply as more and more desirable material finds its way into museums and libraries, often donated by private collectors seeking tax relief.

There are several reasons why one might expect prices for paintings or other types of books to rise more rapidly than those for books on mathematics. For a start, though mathematicians see great beauty in a proof, most people get more aesthetic enjoyment from a Monet, or from a book with a lavish binding or with literary associations. Further, it has been noted that while almost everyone will know about Rembrandt, Bach and Shakespeare and the educated will know about Fabritius, Buxtehude and Marlowe, few educated people could identify Euler, Lagrange and Fermat. Yet these are mathematicians of the stature of the former group instead of the latter. So it might seem that great books and manuscripts in mathematics would be of interest only to collectors who are also mathematicians, and mathematicians are notorious for not caring about their history.

In spite of all of this, prices of rare mathematical material have been climbing steadily over the past 20 years. A case in point is the valuable collection of Robert Honeyman. Mr. Honeyman, a graduate of Lehigh University (1920), laid the foundations for his science and literature collections in the 1930's when he was based in New York. His collections continued to grow after his move to California until the collector was nearly evicted from his house by his collection. The collection of English and American literature was given to Lehigh University in 1959 but the scientific collection continued to grow until purchased by Sotheby Parke Bernet & Co. and sold at auction in a series of seven sales between October 1978 and May 1981. These auctions have set new records for many seldom seen items in the history of mathematics. At this sale, Euler's important *Introductio ad analysin infinitorum* of 1748 (in which he introduced the partition function and in which he evaluated the zeta function at some positive integral values, among other things) was sold for $660, not a surprisingly high price until one realizes that less than 20 years ago a West Coast dealer had two copies of this two volume set on his shelves, each set priced at $75. (It should also be noted that if the copy sold at Sotheby's was purchased by a dealer for resale it will almost certainly be priced at about twice the amount paid, so the collector may end up paying roughly $1200–$1300 for this set, a price in line with recent catalogue prices.) Another great Euler, the *Methodus inveniendi lineas curvas* (1744), viewed by some as his greatest work and often called the first work in the calculus of variations, was sold for $2000. Fifteen years ago this book could be found in catalogues for roughly $500. (All comparisons here are between copies in similar condition.) In a somewhat earlier auction of books from the same collection, the *Opera omnia* (1742) of Johann Bernoulli sold for $2000; less than 20 years ago a London dealer had several copies on his shelves, each marked £28, roughly $75 at that time.

The books just mentioned are important books but they are not books that would have as great appeal to the collector untrained in mathematics as, for instance, the first editions of Copernicus' *De revolutionibus* (1543), Newton's *Principia* (1687), or the first printed Euclid of 1482. These books are great monuments in the history of science, indeed, in the history of Western thought, and the prices reflect this. They are well beyond the means of all but a few collectors. The Copernicus brought $58,000 at the Honeyman sale; one could buy a copy in 1963 for about $900. The Honeyman copy of the 1482 Euclid brought $39,600, but in the early 1940's one could buy this book for around $200. In 1957 a first edition of Newton's *Principia* sold for $250, and in recent years it passed $10,000, then $15,000, and a nice copy was recently offered by a London dealer for $35,000.

Other books, not nearly as important as the Newton or the Copernicus, nevertheless, have recently commanded astonishing prices. Pascal's *Traité du triangle arithmétique* (1665), for example, is an unusual little book but certainly not of great mathematical importance. After all, what we know as Pascal's triangle was known to Nasir-Ad-Din, the Persian, in 1265, according to [4], and was known to the Chinese even earlier, around 1100 [3]. Even in Europe it appeared in a work of Apianus in 1527, so Pascal's treatise certainly is not the first appearance of this array (though, to give Pascal his due, one must admit that he did investigate properties of elements of the array, both as elements defined recursively and as combinations). Experts estimated that this book might bring between $1850 and $2400 when it came up at the Honeyman sale. Instead, it brought $9500. This price might be attributed to the fact that Pascal has a reputation outside mathematics, but even those interested in his literary or theological writing would surely hesitate to pay prices like that for his scientific work.

These examples attest to the fact that many investors feel that books, like art and antiques, are a good hedge against inflation. But not all those who buy old mathematics books are investors—a collector buys books because collecting is fun.

A collector enjoys owning and reading a book that represents the first appearance in print of a great idea. And then there are the dealers, who are often remarkably interesting, well-informed people, so visiting a good bookshop is almost a social occasion. One never knows what will turn up in a shop or in a dealer's catalogue. For example, in the mid 1960's, one of the authors found in an English dealer's catalogue Hardy's copy (signed by Hardy) of Waring's *Meditationes Algebraicae* (1770). This is an interesting book on two counts: it belonged to Hardy who, with Littlewood, worked for many years on Waring's problem, and it contains the first appearance in print of what is now known as Wilson's Theorem (a theorem neither first observed by nor proved by Wilson!). The book cost less than $35.

On another occasion the experience was more expensive: the discovery in San Francisco of the published version of Gauss' doctoral dissertation in which he gave the first generally accepted proof of the fundamental theorem of algebra. For this, the price paid may have been too much or it may have been a bargain, but one may never know because the previous recorded sale of this book (scarcely more than a pamphlet) was in 1928.

One summer in an Oslo bookstore we asked the usual question: Do you have anything interesting in mathematics? The staff assured us that they did not. Then the owner remembered having something downstairs, but he was sure we would not be interested in such a thing. We pressed him and finally he produced a run of nine issues of the Norwegian journal *Magazin for Naturvidenskaberne* for 1823–25. These contained four papers by Abel written when he was a student, only three of which we could find later in the *Oeuvres Complètes*, edited by Holmboe in 1839. That Holmboe's edition was less than complete is not surprising. Abel's great work on elliptic functions, submitted to the Paris Academy in 1826 and lost by Cauchy and Legendre, was not published until 1841, two years after Holmboe's edition. But surely Holmboe would have known all of the early works in the *Magazin for Naturvidenskaberne*, since he contributed to the same journal. This mystery was cleared up only by searching through the 1881 edition of the *Oeuvres Complètes* where the editors, L. Sylow and S. Lie, explain the omission of this paper "dans lequel it s`était glissé, par inadvertance, une faute grave." So the Oslo visit had turned up a paper of Abel's that contained an error and which, Sylow and Lie go on to explain, Abel had, in fact, withdrawn!

The journal turned out to be interesting in other ways. It was founded in Oslo (Christiania) in 1823 to provide an opportunity to Norwegian scientists to publish their work. One of the editors was Hansteen, who contributed numerous articles, some on magnetism and the location of the earth's magnetic poles. Hansteen must have anticipated some objections to the inclusion of Abel's mathematical papers because to two of the papers he adds short apologetic essays, explaining the value of mathematics, presumably to readers who wanted general scientific articles on rocks, birds and physical phenomena. Hansteen says: "It may seem that in a periodical intended for the natural sciences, a memoir in pure mathematics is not in its right place. But mathematics is nature's doctrine of pure form. For the scientist it is similar to the dissecting knife of the anatomist, an absolutely necessary tool without whose aid one cannot penetrate the surface. Over the curtain which hides the entrance to the inner sanctum, the master builder of nature has placed the same motto as the Greek philosopher above the entrance to his lecture hall: 'Let no one ignorant of geometry enter.'" The complete essays (translated, fortunately, from the Norwegian) are included in Ore's biography of Abel [5].

The Oslo bookseller made a sale that day, whether he wanted to or not. The price of the journals was approximately $70.

In perusing early editions, the collector sometimes marvels at the mathematics, and sometimes at the ingenuity of the author, often made more impressive because of inadequate notation. And then again, there are the cases where one marvels at the notation. In the first appearance in print of the differential calculus, Leibniz' "Nova methodus pro maximis et minimis" in the *Acta Eruditorum* for 1684, the notation is the same as that used in the most recent books on calculus.

Sometimes a book affords a glimpse of the personality of the author. For example, following his successful derivation of formulas for the Bernoulli polynomials, the following paragraph occurs in Jacques Bernoulli's *Ars Conjectandi* (1713) (the paragraph is taken from the English translation of 1795 [1]):

> I cannot but observe on this occasion, that the learned Ismael Bullialdus, or Bouillaud, has been rather unfortunate in his manner of treating this subject, in his *Treatise on the Arithmetick of Infinites*; since the whole of the folio volume which he has written upon it does nothing more than enable us to find the sums of the first six powers of the natural numbers, 1, 2, 3, 4, 5, 6, 7, etc. continued to any given number *n*; which is only a part of what we have here accomplished in the compass of a dozen pages.

Bernoulli knew how to put someone in his place.

Sometimes one comes across something of pedagogical interest. Some years ago one of the authors was systematically searching through London bookshops (where, incidentally, he was told in one of the most prestigious, that London is not the place to find early mathematics books—he would have better luck in Los Angeles!). In one of the tiny cluttered shops in Cecil Court, the owner recalled that there was, indeed, a Euclid somewhere in the shop. He went down into the basement and after a considerable time emerged clutching a dusty copy of the 1847 edition of Euclid, redone by Oliver Byrne, "Surveyor of Her Majesty's Settlements in the Falkland Islands." Byrne must have had many long evenings to kill in that remote spot, so he devised a way of teaching geometry using colors. When one line is to equal another, or areas are to be equal, they are given the same color. The book, although published by Pickering in London, was printed by the Chiswick Press and is certainly one of the most spectacular Euclids ever. The colors (black, red, blue and yellow) are flat and extremely vivid—they were laid on by wood block printing—and the paper is as crisp and as clean as if it were printed last week. In the mid-19th century, it was clearly too expensive a process to be used widely, but Byrne anticipated the color printing used in modern texts. Collecting editions of Euclid alone could be a lifetime hobby, from the 1482 incunable mentioned earlier to the first in a modern language (Italian, 1543), the first in English (1570), with the remarkable preface of John Dee [6], all the way up to the elegant edition designed for Random House in 1944 by Bruce Rogers and containing an essay on geometry by Paul Valéry.

Sometimes in searching for good mathematics books one finds something quite different but which one cannot pass up. For example, in a San Francisco bookshop one of the authors turned up a copy of J. J. Sylvester's poem, *Spring's Debut: A Town Idyll*, privately published and, no doubt, at the author's expense. It is corrected in the author's own hand and there is a presentation inscription to a fellow professor at Johns Hopkins. The poem consists of 113 lines—every single line ending in the sound "in", surely one of the more stringent constraints on poetry since the decline of Alexandrine verse! It is an incredible tour de force, but it was the sort of thing that may have led to the terse comment on Sylvester's poetry in the *Dictionary of American Biography*: "Most of Sylvester's original verse showed more ingenuity than poetic feeling." Cajori [2] remarks that "at the reading, at the Peabody Institute in Baltimore, of his [Sylvester's] Rosalind poem, consisting of about 400 lines all rhyming with 'Rosalind', he first read all his explanatory footnotes, so as not to interrupt the poem; these took one hour and a half. Then he read the poem itself to the remnant of his audience."

Those who contemplate collecting mathematics books should be aware of the diminishing supply. The great books of Napier, Kepler, Copernicus, Newton, and Fermat have practically disappeared from the market. (Newton's theological discourses are still to be had at modest cost!) The later classics—Gauss,

Euler, Lagrange, Laplace, and so on—are now priced out of the reach of many. First editions of modern classics are now appearing in dealers' catalogues: Hardy and Wright's *Introduction to the Theory of Numbers*, Pólya and Szegö's *Aufgaben and Lehrsatze*, books by Hilbert and Minkowski. These books sell in the $40–$100 range. These and classic texts, as well as minor books from earlier centuries, may be the standard fare for collectors in the future.

For those interested in seeing the great books in the history of mathematics, visits to any large university rare book room will be worthwhile. There are some collections that are particularly notable, and displays of materials are often scheduled: The Burndy Library in Norwalk, Connecticut; the De Golyer Collection at the University of Oklahoma; the Brasch Isaac Newton Collection at Stanford University; the Stanitz Collection at Kent State University; the Milestones of Science Collection at the Buffalo Society of Natural Sciences; the Burndy Collection at the Smithsonian Institution; the William A. Cole Collection at the University of Wisconsin; the Herbert McLean Evans Collection and the Prandtl Collection at the University of Texas at Austin.

References

1. James Bernoulli, *The Doctrine of Permutations and Combinations, Being an Essential and Fundamental Part of the Doctrine of Chances*, Maseres, London, 1795, p. 197.

2. Florian Cajori, *A History of Mathematics*, 2nd ed., Macmillan, New York, 1919, p. 343.

3. Lam Lay-Yong, The Chinese connection between the Pascal Triangle and the solution of numerical equations of any degree, *Historia Math.*, vol. 7, no. 4, pp. 407–24.

4. Nasir-Ad-Din At-Tusi, Handbook of arithmetic using board and dust, (Russian translation by S. A. Ahmedov and B. A. Rosenfeld) *Istor.-Mat. Issled.*, vol.15, pp. 431–44.

5. Øystein Ore, *Niels Henrik Abel, Mathematician Extraordinary*, University of Minnesota Press, Minneapolis, 1957, pp. 62–63.

6. John L. Thornton and R. I. J. Tully, *Scientific Books*, Libraries and Collectors, 2nd ed., The Library Association, London, 1962, pp. 11–12.

Book prices have not come down since this article was written, nor has the supply of old books increased. I have seen a first edition of Boole's *Laws of Thought* (1854) advertised for sale at $3500, which seems quite a bit for something that only recently saw its sesquicentennial. A second edition of Descartes' *Géométrie* (translated into Latin, 1659) will set you back $4450. Even a 1965 Chelsea reprint of Landau's *Differential and Integral Calculus* fetches $60. To build up a mathematical collection, it is probably best to follow the Alexanderson-Klosinski method and poke around used book stores. Among the dross (and there is plenty of that!) a few gems may be found.

To the list of notable collections of mathematics books should be added the William Marshall Bullitt Collection of Rare Mathematics and Astronomy at the University of Louisville. Mr. Bullitt (1873–1957) was a lawyer, but an enthusiast for mathematics with many mathematicians among his acquaintances. G. H. Hardy caused him to set a goal of gathering first editions of books by the twenty-five greatest mathematicians of all time, but no collector can stop with twenty-five and the collection grew to some 370 volumes.

For the benefit of those who did not recognize the names of Fabritius, Buxtehude, and Marlowe (Fabritius was new to me), Carel Fabritius (1622–1654) was a Dutch painter, Dietrich Buxtehude (1637–1707) was a German organist who J. S. Bach once walked 200 miles to hear play, and Christopher Marlowe (1564–1593) was an English poet and dramatist.

Gerald Alexanderson teaches at Santa Clara University. He has edited *Mathematics Magazine* and has served as both Secretary and President of the Mathematical Association of America. Leonard Klosinski is also at Santa Clara. He is the Director of the William Lowell Putnam Mathematical Competition.

23

Irrational Square Roots

Everyone knows the proof, attributed to Pythagoras, that $\sqrt{2}$, is irrational. I will not repeat it here because it appears at the beginning of the selection that follows. After disposing of $\sqrt{2}$ it is natural to inquire into $\sqrt{3}, \sqrt{5}, \sqrt{6},\ldots$ and the ancient Greeks did. Theodorus, a teacher of Plato, showed that the square roots of the non-square numbers up to 17 were irrational, but there he stopped. Why did he quit there? His method is lost, so we will never know for sure, but that has not kept people from making conjectures.

One, made I think in all seriousness, was that he stopped because a spiral of right triangles (see Figure 1) could be continued no further without overlap. This seems very unlikely.

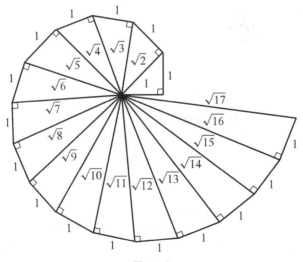

Figure 1

The best that Hardy and Wright could do, in their marvelous *The Theory of Numbers*, was to say, as our author notes, "he may well have been quite tired." This is unworthy of Hardy and Wright, but they may have been tired.

The method in the author's reference [6] is based on the idea that later appears in Euclid's *Elements*, that

> If, when the lesser of two unequal magnitudes is continually subtracted in turn from the greater, that which is left never measures the one before it, the magnitudes will be incommensurable.

(This is the same as saying that a number has a finite continued fraction representation if and only if the number is rational, but the ancient Greeks had not thought of continued fractions.) For example, applied to $\sqrt{5}$ and 1, from

$$\sqrt{5} - 2 \cdot 1 = \sqrt{5} - 2,$$

we see that by continually subtracting 1 (two times) from $\sqrt{5}$ we get a new greater magnitude, 1, and a new lesser one, $\sqrt{5} - 2$. Now let us do it again: since

$$\frac{1}{\sqrt{5} - 2} = \frac{1}{\sqrt{5} - 2} \cdot \frac{\sqrt{5} + 2}{\sqrt{5} + 2} = \sqrt{5} + 2 = 4 + \left(\sqrt{5} - 2\right),$$

we have

$$1 - 4 \cdot \left(\sqrt{5} - 2\right) = \left(\sqrt{5} - 2\right)^2.$$

The greater magnitude is now $\sqrt{5} - 2$ and the lesser one is $\left(\sqrt{5} - 2\right)^2$. Let us do it one more time:

$$\frac{\sqrt{5} - 2}{\left(\sqrt{5} - 2\right)^2} = \frac{1}{\sqrt{5} - 2} = \sqrt{5} + 2 = 4 + \left(\sqrt{5} - 2\right),$$

so $\sqrt{5} - 2 - 4 \cdot \left(\sqrt{5} - 2\right)^2 = \left(\sqrt{5} - 2\right)^3.$

The greater magnitude is now $\left(\sqrt{5} - 2\right)^2$ and the lesser one is $\left(\sqrt{5} - 2\right)^3$. If we continue the process, the greater will be $\left(\sqrt{5} - 2\right)^n$ and the lesser will be $\left(\sqrt{5} - 2\right)^{n+1}$, n without end. Since the process does not terminate, $\sqrt{5}$ and 1 are incommensurable and $\sqrt{5}$ is irrational.

Theodorus *could* have done this (not the algebra, of course, but the process—it is possible to cast it in a geometrical form), but the author of the next selection has convinced me that he didn't. His explanation is so compelling that I am sure that it describes what Theodorus actually did and why he had to stop at $\sqrt{17}$. A historical mystery has been solved, as much as it can be.

Theodorus' Irrationality Proofs

Robert L. McCabe (1976)

The Pythagorean proof of the irrationality of $\sqrt{2}$ is well known. If we assume that $\sqrt{2} = a/b$ and that a/b is in lowest terms, then $2b^2 = a^2$ implies that a^2 and hence that a is even, that is, $a = 2k$. Substituting $2k$ for a yields $2b^2 = 4k^2$, or $b^2 = 2k^2$. Therefore b^2 and hence b is even. But a and b cannot both be even if a/b is in lowest terms. This proof, using only the concepts of even and odd, generalizes to almost all other square roots thereby shedding light on an ancient problem concerning incommensurables, or irrational numbers. Plato's *Theaetetus* contains a famous passage on irrational numbers. Theaetetus says [**1**, p. 25],

> Theodorus here was drawing some figures for us in illustration of roots, showing that squares containing three square feet and five square feet are not commensurable in length with the unit of the foot, and so, selecting each one in its turn *up to* the square containing seventeen square feet; and at that he stopped. [Italics added]

The question that has puzzled historians ever since is this: What manner of proof did Theodorus use? It is certain that Theodorus knew the proof of the irrationality of $\sqrt{2}$, but did not have any general method of proof available, else why stop at 17? Van der Waerden [**6**, pp. 142–145], following Zeuthen, suggests a method of proof involving ratios, which after a few transformations begin to cycle themselves endlessly thus leading to proofs by contradiction. The method works well through 17 (18 is trivial since $\sqrt{18} = 3\sqrt{2}$), and 19 is "quite complicated" because it requires eight ratios before the endless cycle begins. Van der Waerden may be right, but his suggestion is not convincing.

Hardy and Wright [**2**, pp. 41–43] suggest the familiar method involving remainders. For example, if $\sqrt{5} = a/b$ with a/b in lowest terms, then $5b^2 = a^2$ implies that 5 divides a^2. Does 5 then divide a? There are five possibilities: a is of the form $5n$, $5n + 1$, $5n + 2$, $5n + 3$, or $5n + 4$. Squaring each shows that only $(5n)^2$ turns out to be divisible by 5. Thus $a = 5n$. Substituting, we easily find that 5 also divides b so that a/b was not in lowest terms. Unfortunately this proof works for all squares and Hardy and Wright can give no explanation for Theodorus stopping at 17 except that "he may well have been quite tired."

Heath [**3**, p. 133] suggests that Theodorus "may have adapted the Pythagorean proof in the case of $\sqrt{2}$, by substituting 3, 5,... for 2." This suggestion is
quite perceptive, for the adaptation of the famous even-odd Pythagorean proof
fails precisely at 17—a remarkable fact that has apparently gone unnoticed.
(Prof. Dirk Struik has called my attention to an unpublished Ph. D. dissertation
[**4**] of W. R. Knorr in which a similar conclusion is reached. Knorr suggests
a rather long detour through Pythagorean triples, but in essence argues for an
even-odd mode of proof.) The following theorem, also proved by Knorr, sheds
light on the matter and may well supply the answer to Theodorus' method.

THEOREM. If p is a positive integer which can be written in any one of the following forms, $4n + 2$, $4n + 3$, or $8n + 5$, for $n = 0, 1, 2, \ldots$, then $\sqrt{p}$ can be
proved irrational using only even-odd techniques.

The proof is easy. Assuming, for example, that $\sqrt{8n+5} = a/b$ in lowest
terms, we have $(8n + 5)b^2 = a^2$, so that both b and a must be odd. Letting $b =
2j + 1$ and $a = 2k + 1$, substituting and simplifying, we get the equation
$$8nj^2 + 8nj + 2n + 5(j^2 + j) + 1 = k^2 + k. \tag{1}$$
Since the combinations $j^2 + j$ and $k^2 + k$ are always even, for all j and k, we
have a contradiction—for now the left side of (1) is odd and the right side even.
The proofs for $4n + 2$ and $4n + 3$ are similar. Thus only numbers of the form
$4n + 1$ are not covered by the theorem (those of the form $4n$ are easily reduced
to one of the forms since $\sqrt{4n} = 2\sqrt{n}$). But, since $8n + 5 = 4(2n + 1) + 1$, it is
only numbers of the form $4(2n) + 1 = 8n + 1$ whose square roots cannot easily
be handled by this method. These numbers are
$$1, 9, 17, 25, 33, 41, 49, 56, \ldots.$$
The list contains all the odd perfect squares because they are all of the
form
$$8\left(\frac{x(x+1)}{2}\right) + 1, \quad \text{where } x = 0, 1, 2, \ldots. \tag{2}$$
Therefore the theorem excludes precisely those positive integers *not* of the
form (2).

Thus a direct application of the Pythagorean even-odd technique fails precisely at 17. A straightforward attack on 17, using only even and odd techniques, proves fruitless, or at least too tedious to warrant continuing. The first
step is to assume $\sqrt{17} = a/b$ with a/b in lowest terms. Letting $b = 2j + 1$ and $a
= 2k + 1$ leads to the equation
$$17(j^2 + j) + 4 = k^2 + k. \tag{3}$$
There is no even-odd contradiction here, hence there are four possibilities: j
and k are both even, j is even and k is odd, etc. Trying each in turn leads again

to still further possibilities. The reader is encouraged to try it.

The coincidence here seems too strong to resist. Knowing the Greeks' powerful belief in the theory of the even and the odd, it seems reasonable that Theodorus may well have tried using this theory on the general problem of irrationals.

The translation of Plato is critical. Referring to the quotation from the *Theaetetus*, van der Waerden [6, p. 142] translates the latter part of

$$... \kappa\alpha\iota \ \sigma\acute{v}\tau\omega \ \kappa\alpha\tau\alpha \ \mu\acute{\iota}\alpha\nu \ \epsilon\kappa\acute{\alpha}\sigma\tau\eta\nu \ \pi\rho\omicron\alpha\iota\rho\omicron\acute{v}\mu\epsilon\nu\omicron\sigma \ \mu\acute{\epsilon}\chi\rho\iota$$
$$\tau\eta\zeta \ \epsilon\pi\tau\alpha\kappa\alpha\iota\delta\epsilon\kappa\alpha\pi\omicron\delta\omicron\zeta \ - \ \epsilon\nu \ \tau\alpha\acute{v}\tau\eta \ \pi\omega\zeta \ \epsilon\nu\acute{\epsilon}\omicron\chi\epsilon\tau\omicron$$

as

... up to the one of ′17 feet; here something stopped him (or: here he stopped).

Heath [3, p. 132] writes

... up to seventeen square feet, 'at which point for some reason he stopped'.

The Greek word $\mu\epsilon\chi\rho\iota$ meaning "up to" or "until" has more the sense of "just short of" [5, p. 1123] and suggests what I believe to be true: Theodorus generalized the Pythagorean even-odd concept and, as shown by the theorem, got stuck at 17. He did not have the algebra necessary to prove the theorem in its full generality and would have handled each integer separately. It is worth quoting van der Waerden again [6, p. 109]:

For the Pythagoreans, even and odd are not only the fundamental concepts of arithmetic, but indeed the basic principles of all nature.

And,

Plato always defines Arithmetica as 'the theory of the even and the odd'.

References

1. H. N. Fowler, *Plato: Theaetetus—Sophist*, Cambridge, Mass., 1921.

2. G. H. Hardy and E. M. Wright, *An Introduction to the Theory of Numbers*, 4th ed., London, 1960.

3. T. L. Heath, *A Manual of Greek Mathematics*, London, 1931.

4. W. R. Knorr, The Pre-Euclidean Theory of Incommensurable Magnitudes, unpublished doctoral dissertation, Dept. of the History of Science, Harvard University, 1972.

5. H. G. Liddell and R. Scott, *A Greek-English Lexicon*, revised and augmented by H. S. Jones, Oxford, 1968.

6. B. L. van der Waerden, *Science Awakening*, Groningen, Holland (English translation by Arnold Dresden), 1954.

Robert L. McCabe received his Ph.D. degree from Boston University in 1971 and at the time this paper was published was teaching at Southeastern Massachusetts University.

Wilbur Knorr, whose 1972 dissertation contained the explanation given in the paper, went on to a highly distinguished career as a historian of mathematics. He died young, at the age of 51, in 1997.

24

The Euler-Diderot Anecdote

There really is a shortage of good mathematical anecdotes. Why this is so is not clear, though I suspect that a good part of the reason is that mathematicians' exterior lives tend to be rather dull. Their interior dramas are quite enough for them. Think of Gauss, never leaving Göttingen, going to the library for an hour or two every morning to read newspapers—not the stuff of anecdote, and I cannot at the moment recall any punchy Gauss anecdotes. There may be none.

The shortage is, and has been, so acute that anecdotes have been *made up*. Once fabricated, some of them get repeated often enough so that they pass into the canonical history of mathematics. They can do so even when they are *bad* anecdotes, such as the famous Euler-Diderot anecdote, repeated several times below. The E-D anecdote is false, is on the face of it not credible, and misrepresents the characters of both Euler and Diderot. *Still* it gets repeated. It must be that the hunger for anecdotes is so great that people will swallow ones that, if there were a better supply available, would make them gag. That is what starvation can do.

I hope that no one reading the following will repeat the Euler-Diderot anecdote ever again. It may then die out. Anecdotes are tough to kill, but, after enough centuries go by, we may see it disappear.

The So-called Euler-Diderot Incident

R. J. Gillings (1954)

1. Introductory Notes. Peter I or Peter the Great, the first emperor of Russia, died in 1725 and was succeeded by his wife Catherine I. She reigned only two years, being followed by Peter II who died in 1730. Then came in relatively quick succession, Anne, daughter of Peter's half-brother, 1730–40; Ivan VI, grand-nephew of Anne, 1740–41; Elizabeth, daughter of Peter and Catherine I, 1741–61; Peter III, grandson of Peter, 1761–62; and Catherine II who reigned the 34 years from 1762 to 1796.

Catherine II is by some historians referred to as Catherine the Great, while others deny her the right to that title. However she was a woman of culture who respected the arts generally, encouraged men of letters, and who herself displayed ability as a writer.

Denis Diderot (1713–84) was a distinguished philosopher, an encyclopaedist, and an author of many scientific publications, who visited St. Petersburg in 1773–74 since the Empress Catherine II had purchased his library and appointed him its first librarian. The following mathematical memoirs of Diderot published in 1784, among others, are discussed by L. G. Krakuer and R. L. Krueger, [11: "Sur la tension des cordes," "De la développante du cercle," "Résistance de l'air au mouvements des pendules"; and in J. L. Coolidge's, *The Mathematics of Great Amateurs*, Oxford, 1949, ch. XIV, Diderot's treatment of "vibrating strings" is discussed, together with those mentioned above. Diderot, who had been in turn, a deist, a pantheist, a sceptic and finally an atheist, was then verging on 60 years of age, Catherine was 44 and Euler was 66.

Leonard Euler (1707–83), famous mathematician, was invited to St. Petersburg in 1727, to accept the chair of mathematics at the Academy recently founded by Peter the Great. Here he remained until 1741, when he was induced by Frederic the Great of Prussia to spend 25 years in Berlin until 1766, when he returned to St. Petersburg at the request of Catherine II, and worked there until the end of his life.

Dieudonné Thiébault (1733–1807) was a French man of letters who for the period from 1765 to 1784, lived in Berlin at the invitation of Frederic the Great. He was a member of the Berlin Royal Academy. The incident to which this paper refers had its origin in Thiébault's reminiscences of his twenty years sojourn in Berlin.

2. De Morgan's Statement [2]. "The following anecdote is found in Thiébault's *Souvenirs de vingt ans de séjour à Berlin*, published in 1804. Thiébault does not claim personal knowledge of the anecdote, but he vouches for its being received as true all over the north of Europe. Diderot paid a visit to Russia at the invitation of Catherine the Second. At that time he was an atheist or at least talked atheism: it would be easy to prove him one thing or the other from his writings. His lively sallies on this subject much amused the Empress, and all the younger part of her Court. But some of the older courtiers suggested that it was hardly prudent to allow such unreserved exhibitions. The Empress thought so too, but did not like to muzzle her guest by an express prohibition: so a plot was contrived. The scorner was informed that an eminent mathematician had an algebraical proof of the existence of God, which he would communicate before the whole Court, if agreeable. Diderot gladly consented. The mathematician, who is not named, was Euler. He came to Diderot with the gravest air, and in a tone of perfect conviction said, '*Monsieur!* $(a + b^n)/n = x$, *donc Dieu existe; répondez!*' Diderot, to whom algebra was Hebrew, though this is expressed in a very roundabout way by Thiébault—and whom we may suppose to have expected some verbal argument of alleged algebraical closeness, was disconcerted, while peals of laughter sounded on all sides. Next day he asked permission to return to France, which was granted."

3. Cajori's Statement [31]. "The story goes that when the French philosopher Denis Diderot paid a visit to the Russian Court, he conversed very freely and gave the younger members of the Court circle a good deal of lively atheism. Thereupon Diderot was informed that a learned mathematician was in possession of an algebraical demonstration of the existence of God, and would give it to him before all Court, if he desired to hear it. Diderot consented. Then Euler advanced towards Diderot, and said gravely, and in a tone of perfect conviction: '*Monsieur*, $(a + b^n)/n = x$, *donc Dieu existe; répondez!*' Diderot, to whom algebra was Hebrew, was embarrassed and disconcerted, while peals of laughter rose on all sides. He asked permission to return to France at once, which was granted." [A reference is then given to De Morgan's *Budget of Paradoxes*.]

4. Bell's Statement [4]. "We shall tell once more the famous story of Euler and the atheistic (or perhaps only pantheistic) French philosopher Denis Diderot (1713–1784). Invited by Catherine the Great to visit her Court, Diderot earned his keep by trying to convert the courtiers to atheism. Fed up, Catherine commissioned Euler to muzzle the windy philosopher. This was easy because all mathematics was Chinese to Diderot. De Morgan tells what happened (in his classic *Budget of Paradoxes*, 1872): Diderot was informed that a learned mathematician was in possession of an algebraical demonstration of the existence of God, and would give it before all the Court, if he desired to hear it. Diderot gladly consented.... Euler advanced towards Diderot, and said gravely, and in a tone of perfect conviction: 'Sir, $(a + b^n)/n = x$, hence God exists; reply!' It sounded like sense to Diderot. Humiliated by the unrestrained laughter which greeted his embarrassed silence, the poor man asked Catherine's permission to return at once to France. She graciously gave it. Not content with this masterpiece, Euler in all seriousness painted his lily with solemn proofs, in deadly earnest, that God exists and that the soul is not a material substance. It is reported that both proofs passed into the treatises on theology of his day."

5. *Hogben's Statement* [5].

"There is a story about Diderot, the Encyclopaedist, and materialist, a foremost figure in the intellectual awakening which immediately preceded the French Revolution. Diderot was staying at the Russian court, where his elegant flippancy was entertaining the nobility. Fearing that the faith of her retainers was at stake, the Tsaritsa commissioned Euler, the most distinguished mathematician of the time, to debate with Diderot in public. Diderot was informed that a mathematician has established a proof of the existence of God. He was summoned to court without being told the name of his opponent. Before the assembled court, Euler accosted him with the following pronouncement, which was uttered with due gravity: '$(a + b^n)/n = x$, *donc Dieu existe, répondez*!' Algebra was Arabic to Diderot. Unfortunately he did not realise that was the trouble.... Translated freely into English it may be rendered: 'A number x can be got by first adding a number a to a number b multiplied by itself a certain number of times, and then dividing the whole by the number of b's multiplied together....' Like many of us Diderot had stagefright when confronted by a sentence in sign language. He left the court abruptly amid the titters of the assembly, confined himself to his chambers, demanded a safe conduct, and promptly returned to France."

6. *Thiébault's Statement* [6].

An English translation of the relevant passage is as follows: From the moment of his arrival Diderot was well received, all his expenses had been paid by the Empress whom he amused immensely by the fecundity and fire of his imagination, by the abundance and singularity of his ideas, and by the zeal, boldness and eloquence with which he publicly upheld atheism. But several of the older courtiers more experienced and perhaps more easily alarmed, persuaded their autocratic sovereign that teachings of this kind could have unfortunate consequences for the whole court, and especially among the large youthful group, destined for important empire posts, who might embrace this doctrine with more eagerness than careful scrutiny. The Empress then desired that some restraint be put upon Diderot on this subject, provided that she did not appear to play any part in the matter, and provided that no one should show any undue authority about it. It was therefore announced to the French philosopher one evening, that a Russian philosopher, a learned mathematician and a distinguished member of the Academy, was prepared to prove the existence of God to him, algebraically, and before the whole court. Diderot said that he would be happy to listen to such a demonstration, in the validity of which of course, he did not believe, and so an hour and a day were fixed to convince him. The occasion having arrived, with the whole court present, that is to say, the men and more particularly the younger members, the Russian philosopher gravely advanced towards the French philosopher, and speaking in a tone of voice to imply his full conviction, said, "*Monsieur, (a + b^n)/z = x*, [7] *therefore God exists: answer that!*" Diderot was willing to show the futility and stupidity of this so-called proof, but felt in spite of himself, the embarrassment that one would, on discovering, (among them), their intention of making a game of it, so that he was not disposed to attempt to admonish them for the indignities proposed for him. This adventure made him fearful that there might be others in store for him of a like nature, and so sometime afterwards he expressed his desire to return to France. Then the Empress having declared her willingness to pay all his traveling expenses, he was sent on his journey after having received 50,000 francs. Eventually his carriage was wrecked near Riga, but he received from the governor of that town the whole of the cost of the repairs. I do not assert the truth of any one of these facts; I say only, that at that time, they were talked about, and were believed by the inhabitants of the north.

7. *Conclusions.*

Since Thiébault's statement is the only authority for the facts discussed in this incident we may now summarize some of the unwarranted changes made by the authors mentioned. We see that De Morgan's statement makes more than one departure from his quoted authority: "Algebra was Hebrew" ... (see also the facts given in the Introductory Notes), "Diderot ... was disconcerted while peals of laughter sounded on all sides," "Next day he asked permission to return to France," and "The mathematician who was not named was Euler." We grant that Thiébault's phrase, "a Russian philosopher, a learned mathematician and a distinguished member of the academy," seems rather definitely to refer to the Swiss Mathematician Euler.

De Morgan's inventions are naturally repeated by Cajori. Bell also repeats them but substitutes "All mathematics was Chinese to Diderot," for "Algebra was Hebrew to Diderot." No authority is given for Bell's statements in his final two sentences, hence we question the adequacy of that authority.

It will be observed that Hogben also alters the form of De Morgan's inventions. Struik has well pointed out [8] the incongruity of the story, which is completely out of character both for the devout Euler and the highly intelligent Diderot. Thiébault himself was not convinced of the truth of it. The extent to which legendary stories of history may be distorted is well illustrated by the so-called Euler-Diderot incident.

References

1. L. G. Krakuer and R. L. Krueger, *Isis*, vol. 33, 1941, p. 219–231.

2. A. De Morgan, *A Budget of Paradoxes*. London, 1872, p. 250 and p. 474. The story appears twice.

3. F. Cajori, *A History of Mathematics*. New York, 1919, p. 233. Second edition revised and enlarged.

4. E. T. Bell, *Men of Mathematics*. New York, 1937, p. 146–147.

5. L. Hogben, *Mathematics for the Million*. New York, 1951, p. 17. Opening paragraph of chapter I.

6. D. Thiébault, *Mes Souvenirs de Vingi Ans de Séjour à Berlin*. Paris, 1804, 3 vols. The passage is on page 141 of volume 3. The English translation published in 1806 in 2 volumes at Philadelphia, under the title of, *Anecdotes of Frederick the Great of Prussia*, makes no mention of this story.

7. The formula is printed in italics and the z of the denominator could possibly be mistaken for a 2. It is however "z" clearly enough and not "n" as De Morgan miscopied it and following him, Cajori, Bell, Smith, Sanford, Hogben and others.

8. D. J. Struik, *A Concise History of Mathematics*. New York, 1948, p. 182.

Richard J. Gillings was an Australian historian of mathematics, specializing in ancient Egypt. His best-known work is *Mathematics in the Time of the Pharaohs* (MIT Press, 1972, reprinted by Dover, 1982).

25

Mathematics Made Difficult

In 1910, Silvanus P. Thompson (1851–1916) wrote *Calculus Made Easy*, with subtitle

> Being a very-simplest introduction to those beautiful methods of reckoning which are generally called by the terrifying names of the differential calculus and the integral calculus.

It was not the first of the books whose authors claim to make the learning of mathematics a thing of ease, nor was it the last. On the market today are *How to Ace Calculus, The Complete Idiot's Guide to Calculus*, and similar works. Thompson's book has been, though, one of the most successful, having never been out of print. The latest edition, revised by Martin Gardner, appeared in 1998 and in 2003 was number 4,281 on Amazon's list of best sellers. Newton's *Mathematical Principles of Natural Philosophy* was in 159,121st place.

Thompson and the other authors who claim, explicitly or otherwise, that they have succeeded where so many others have failed and taken the difficulty out of mathematics fly in the face of one of the greatest of them all, Euclid. The story goes that a member of the ruling class in Alexandria wanted to learn geometry and, naturally wanting the best, asked Euclid to teach it to him. Euclid outlined the course of study, which struck the aristocrat as unduly long and difficult, so he asked for an easier way. "There is," Euclid was supposed to have said, "no royal road to geometry."

It is not necessary to believe that any such encounter between Euclid and a Hellenic prince ever took place, but the truth expressed is independent of who said it: learning mathematics is hard, except for that fraction of humanity for whom it is not, and there is no way around it.

Mathematicians know this, and it can be irksome to them to think that those who buy the mathematics-made-simple books are deluded into thinking otherwise. This may be one of the reasons why Carl Linderholm was moved to write *Mathematics Made Difficult* (World Publishing, 1972, out of print). Another was to be funny. The dust jacket says

> Dr. Carl Linderholm was born in Baton Rouge, Louisiana. He studied at the University of Chicago, is married, and has four children. Before lecturing at Roosevelt University and at the Universities of Michigan and Illinois, he was briefly a waiter but was fired because he could not remember on which side of the plate to put the knife, fork, and spoon.

Later the dust jacket falls into the manner of dust jackets of funny books:

> As to likes and dislikes, his favorite wild insect is the tarantula; favorite tame insect, the cheese mite; favorite instrument, the Acadian accordion; most disliked tame plant, privet; most disliked wild plant, scrub oak; most disliked part of speech, the adverb; most disliked country, Mordor.

Nevertheless, the book *is* funny, at least for those who find funny the sort of thing that is in it. There is no disputing with taste, and the remark "That's not funny" cannot be refuted, but those with any appreciation of academic humor will be amused by Dr. Linderholm's book. There follow two excerpts, (though I am tempted to include many more). The first is devoted to showing that $(x + y)(x - y) = x^2 - y^2$, not in the way that you or I would do it:

$$(x + y)(x - y) = x(x - y) + y(x - y) = x^2 - xy + yx - y^2 = x^2 - y^2$$

but by making it difficult. The second contains good advice on how to deal with those "What's the next number in the sequence?" questions.

Mathematics Made Difficult

Carl E. Linderholm (1972)

Polynomial identities

One of the most basic identities for practical purposes in connection with computations is

$$(X - Y)(X + Y) = X^2 - Y^2.$$

This is sometimes expressed in words: the sum times the difference is the difference of the squares. Since we know how to multiply polynomials (previous section), this should not offer any great difficulty. Every term in the polynomial $(X - Y)$ has degree 1; in other words, the value of the function at an element of the monoid of pure monomials is 0, unless the element of the monoid of pure monomials is $X^h Y^I$ with $h + I = 1$. Note that for simplicity it has been taken for granted that all the polynomials are polynomials in the letters X and Y, and that if this supposition is not made then the last statement must be modified. Similarly, the polynomial $(X + Y)$ is a function taking the value 0 on the pure monomial $X^j Y^k$ unless $j + k = 1$. Note further that we have no idea what is meant by the statement "the function takes the value 0 at a monomial", since we have no idea what this 0 is and are even unaware what ring 0 is the zero element of; and note also that this makes no difference to the argument. Putting

$$h + j = m$$
$$I + k = n,$$

we see that the product of the two polynomials has degree 2; that is, their product is a function that must take the value 0 at $X^m Y^n$ except in case $m + n = 2$. This is because

$$m + n = (h + j) + (I + k) = (h + I) + (j + k) = 2$$

whenever even one term in the convolution sum is non-zero. Hence we know that the product takes the form

$$aX^2 + bXY + cY^2,$$

and it remains only to compute the actual values of a, b, and c. In order to compute a, it is necessary to observe only that if

$$\begin{pmatrix} 1 & 0 & 1 & 0 \\ 0 & 1 & 0 & 1 \\ 1 & 1 & 0 & 0 \\ 0 & 0 & 1 & 1 \end{pmatrix} \begin{pmatrix} h \\ I \\ j \\ k \end{pmatrix} = \begin{pmatrix} 2 \\ 0 \\ 1 \\ 1 \end{pmatrix}$$

then (h, I, j, k) must differ from $(1, 0, 1, 0)$ by an element of the kernel of the homomorphism $Z^4 \to Z^4$ determined by the matrix; hence is of the form

$$(1 + w, -w, 1 - w, +w)$$

where w is an integer. Since it is further required that h, I, j, k be natural numbers, we must necessarily have

$$w = 0.$$

Hence the coefficient of X^2 is the product of the coefficient of X in $(X - Y)$ and the coefficient of X in $(X + Y)$, or $1 \cdot 1$, giving

$$a = 1.$$

In a similar way, if $X^h Y^i$ and $X^j Y^k$ are to make a contribution to the coefficient b of XY, then we must have

$$(h, I, j, k) = (1 + w, -w, -w, 1 + w).$$

There are seen to be two solutions in naturals, namely $(1,0,0,1)$ and $(0,1,1,0)$, and we get $b = 0$. Similarly $c = -1$ and we are done.

Note that the above product $X^2 - Y^2$ becomes $X^2 + Y^2$ over rings of characteristic 2, where $-1 = 1$.

An exactly similar method may be used to show that many other familiar polynomial identities hold. Alternatively, of course, one may verify that the monoid algebra $r[S]$ satisfies the ring axioms; i.e., one may verify that the convolution product is associative:

$$\sum_{v=\iota\mu}\left(\sum_{\iota=\kappa\lambda} p(\kappa)q(\lambda)\right)r(\mu) = \sum_{\kappa\iota=v} p(\kappa)\left(\sum_{\lambda\mu=\iota} q(\lambda)r(\mu)\right)$$

and that the "constant polynomial" taking the value 1 on the neutral monomial and taking the value 0 elsewhere is neutral for convolution. Distributivity over addition must also be verified.

Guess the next number

A great deal of what we learn at school is of little use in later life. This is especially true of mathematics. Beyond the most basic arithmetic, which does have a use in checking the bill in a restaurant, there is very little that is ever used again except by specialists. A knowledge of probability theory is handy for an undertaker, so that he can work out when his customers are likely to need him; a little topological group representation theory is not amiss if you happen to end up a quantum mechanic, repairing other peoples' quanta when they begin to wear out. But for most of us, most of our mathematics molders away slowly as the brain cells blink out, cell after cell, in our heads. It never gets used.

Number guessing is an exception. Why is it not taught in the schools? This is one branch of quasi-mathematical trickery that everybody needs desperately, yet it is never found in the school texts. By number guessing, I mean being able to answer those little riddles, like the following.

Here is a little test. Do not be afraid of it; the questions are all of the same type. Though they start out easy at first, pretty soon they are going to get much, much harder. So keep calm. All you have to do is guess the next number in the sequence. We give you, gratis, some of the numbers, which shows how very kind we are, because then we only ask for one number back from you. Sounds simple, doesn't it? In case you still haven't caught on, the first example is worked for you. Here goes:

(1)   8, 75, 3, 9, ___ .

Now all you have to do is look at the numbers, and then in the blank provided write in the number that seems to you logically ought to go there. Now read the numbers again: eight, seventy-five, three, nine. .... . What was that you were about to say? Was it 17? Right! The only number any sensible person would put there is 17. So we write in the number 17 in the space provided, like this:

(1)   8, 75, 3, 9, <u>17</u>.

That's all you have to do. Good luck!

Everyone has encountered one of these little tests, and we all know how much depends on them—that place in a really good university, that step up in the firm that has been hanging fire for years, that membership of Mensa—the chance to look down on more and more stupid people who cannot guess the next number.

Really, it is inexcusable that this art is not taught in every school. The scientific fact is universally acknowledged that only intelligent people can do these puzzles; moreover, nobody denies that there is a crying need for intelligence in all areas of the national economy. Hence, and one would think the inference would be obvious to any person who can guess the next number even in the easy example we saw just above, all that needs to be done in order to cure a vast proportion of the world's ills is to teach everyone to guess the next number. Because then, naturally, everyone would be intelligent. The only case of which the writer knows in which so simple a remedy is known to exist for so serious a social disease is that of mental illness. It is universally accepted that by examining a man's handwriting, trained graphologists can determine what is his mental character, and in particular diagnose any tendencies to violence, mania, etc. Yet nothing is done to teach sane handwriting! But, lest we deceive ourselves, that might involve some difficulty. Good handwriting is not acquired overnight. Fortunately, the ability to answer the sort of puzzle that is the subject of this section can be acquired in anything from a few seconds to an hour.

There is only one little snag. The people who set the intelligence tests are a very special breed. After you learn how to answer the sort of question we are considering here, you will be absolutely certain that the answer you put down is correct; not so the fellow who made up the test. He has his own way of looking at things. The fact is, by his own intelligence test he may not be very intelligent. He may not recognize a correct answer when he sees one. Perhaps that is why he went into his particular line of work. Who ever thinks to question the intelligence of a man whose very job is testing intelligence? What occupation can you think of that makes you utterly safe from the prying doubter who asks, "Has he got a high I. Q.?" *Quis examinabit examinatores ipsos*?

QUESTION 18. Whether there is a simple method, whereby one may always give a logical answer to the sort of puzzle that says, "Here are some numbers; what number comes next?"; and whether it is easily, painlessly, and rapidly learnt?

Objection 1. It would be very surprising if such a method existed. The example mentioned earlier was 8, 75, 3, 9, ___. There does not appear to be any relationship at all connecting those four numbers. Now if one were a better mathematician, or had a better mind for figures, granted one might be able to see how to connect them up; and then one could see what number follows logically after the ones already given. But surely this subject is simple only to the most accomplished number-cruncher.

Reply to Objection 1. It is true that some knowledge of mathematics is necessary before one can actually write down a "relationship" that will produce these numbers; but it is not all that stupendous a task. Writing down such a relationship is a mere technical difficulty, and if one is good at mathematics he can master the problem of writing down such a relationship in a few minutes. Above all, though, one must not forget that no intelligence test or puzzle of this sort actually requires anyone to write down a mathematical formula. It is not necessary even to understand what it is that connects the numbers and makes them come out in that order. The trick which is going to be explained gives you an easy, one might almost say magic method for giving the right answer *without understanding anything*!

Objection 2. The idea that such questions can be answered easily is absurd in the extreme. Great care is taken in the devising and arranging of the questions to test the examinee's abstract pattern-recognition level. While it is conceivable that an examinee will do slightly better or slightly worse than he ought to do, it is a matter of experience that examinees tend to come out about the same if tested again. Stupid people absolutely cannot do well on this test, except by a once-in-a-million chance, like guessing all the right numbers on the lottery.

Reply to Objection 2. It is true that very stupid people cannot do this test; nor can very uneducated people. It is necessary to be able to write a number, which illiterates and cretins cannot do. Moreover, it is necessary to be able to understand that one ought to write a number, and where. These are not very taxing demands; all the abstract pattern-recognition that is needful is contained in them. One of the abstract patterns that does, admittedly, have to be recognized is the line on which one is meant to write the correct answer.

I answer that there is an easy, painless, and simple method for always writing in a correct answer to one of these brain-twisters. The following are some examples of the kind of thing one gets.

(1) 5, 4, 3, 2, ___ .
(2) 2, 4, 6, 8, ___ .
(3) 1, 3, 5, 7, ___ .
(4) 1, 2, 4, 8, 16, ___ .

And now here are some of the answers one is likely to get, with the reasoning behind them.

Example 1. In order to see more clearly the relationship among the numbers, let us draw a graph (Figure 1).

We see that the dots are on a straight line. It is logical to continue the line, and if we do so, we see that the next dot should be at the place marked with a circle. (This technique is called *extrapolation*; it is also used to determine the price of tobacco in the year AD 2000, and to predict the future fortunes of transcendental meditation among humanoids.) Thus the answer is

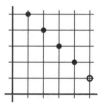

Figure 1

$$5, 4, 3, 2, 1.$$

Example 4. This time, if the numbers are plotted on a graph, we do not get a straight line. Trying to think of something else to do, we first take the logarithms to the base 2, and then plot the graph. These are: $\log_2 1 = 0$, $\log_2 2 = 1$, $\log_2 4 = 2$, $\log_2 8 = 3$, $\log_2 16 = 4$. Hence the graph will be as shown in Figure 2: a straight line; the extrapolation gives 5, and taking 2^5 we get 32. Thus the answer is

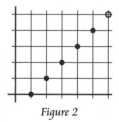

Figure 2

$$1, 2, 4, 8, 16, \underline{32}.$$

Emphatically, this is *not* the method being presented here. Not everyone can get the correct answer by this sort of argument. Before explaining the sensible approach to the problem, it was thought useful to remind the reader how he is *expected* to do the problem. This approach is not systematic. Who could guess that one would get a straight line by taking the logarithms of the numbers in the fourth example? Why not take the exponential function of the numbers, or the inverse tangent?

One easily proves by induction that if f is a polynomial function of degree $\leq n$, and if $f(x_i) = 0$ for every integer I such that $0 \leq I \leq n$ whenever $(x_i)_{0 \leq I \leq n}$ is a finite sequence of real numbers indexed by the set of integers I such that $0 \leq I \leq n$ with the property that $x_i \leq x_j$ if $i \leq j$, then f is identically zero. Moreover, the polynomial function

$$\sum_{k=0}^{n} y_k \frac{\prod_{i \neq k}(x - x_i)}{\prod_{i \neq k}(x_k - x_i)}$$

(under the same hypotheses about (x_i)) has degree $\leq n$, and sends $x_i \to y_i$, for $0 \leq I \leq n$; it must be the unique polynomial that does so. This method provides a systematic method for solving our problem, which gives the formula

$$1 + \tfrac{7}{12}x + \tfrac{11}{24}x^2 - \tfrac{1}{12}x^3 + \tfrac{1}{24}x^4$$

for example 4. Thus if $x = 0$, then we get $1 + 0 + 0 + 0 + 0 = 1$; if $x = 1$ we get

$$1 + \tfrac{7}{12} + \tfrac{11}{24} - \tfrac{1}{12} + \tfrac{1}{24} = 2;$$

if $x = 2$ we get

$$1 + \tfrac{14}{12} + \tfrac{44}{24} - \tfrac{8}{12} + \tfrac{16}{24} = 4;$$

if $x = 3$ we get

$$1 + \tfrac{21}{12} + \tfrac{99}{24} - \tfrac{27}{12} + \tfrac{81}{24} = 8;$$

and so on.

If we apply this formula in the case $x = 5$, we get the answer

$$1 + \tfrac{35}{12} + \tfrac{275}{24} - \tfrac{125}{12} + \tfrac{625}{24} = 31.$$

It may be noted that by logarithms, as the example is often done, we got not 31, but 32. Does this worry us? Not a bit! First of all, there is not much difference between 31 and 32; the difference is only $32 - 31 = 1$. Secondly, even if we are forced, in an unguarded moment, to admit that 31 and 32 are not quite the same, there remains the question: which answer is better anyhow,

1, 2, 4, 8, 16, <u>31</u>, or

1, 2, 4, 8, 16, <u>32</u> ?

Which of these is really more logical, more true to a mathematical way of thinking? Which really shows the greater pattern-recognition facility? There can be no doubt that 31 is the better answer by a long-to-middling chalk; and this for the simplest of reasons. It must be admitted that if anyone should be so narrow-minded and curious as to put down 32 as the number that follows 1, 2, 4, 8, 16, then he had some reason for doing so. It is possible that he arrived at 32 in any of several ways. He may have noticed immediately that the numbers 1, 2, 4, 8, 16, if exchanged for the letters of the alphabet that correspond to them, are just the initial letters of the words "Alien birds do have peculiar ... "; and that these are the words of an old Kentish proverb, the last word "feathers" being omitted. Arranging the alphabet in a circle with z next to a, and counting on past z to a, which then becomes 27, we see that the letter f, with which the extra word "feathers" begins, gets the value 32, which is the right answer. This is all well and good, and is a reasonable interpretation of the problem showing remarkable skill at pattern-recognition. But one can argue, of course, that the letter w, by its very name, is not really a letter at all; it is merely a way of writing the letter u, or just perhaps the letter v, when they are doubled. Hence this letter must be omitted in the counting, and while 32 is a good approximation to the correct answer, it is not as good as the true answer 31.

But it is also possible that the answer 32 was obtained in another way. It could even have been obtained by the graphical method explained earlier, in which by lucky chance someone thought of taking the logarithm. This leads to the formula

$$a_n = 2^n,$$

where of course the first number of the sequence is considered as the zeroth or noughtth number of the sequence. On the other hand, our own work, which was based on a system, gives

$$a_n = 1 + \tfrac{7}{12} n + \tfrac{11}{24} n^2 - \tfrac{1}{12} n^3 + \tfrac{1}{24} n^4.$$

Now both of these answers are backed up by perfectly good mathematical formulae, and so both are perfectly logical. Of the two answers, which is the better—both being correct? The answer is, of course, that the second answer is to be preferred, because it is much the simpler, and is easier to use, and is obtained by a more general method.

On the other hand, some people may prefer the answer 32 and the formula $a_n = 2^n$, because here we have a homomorphism from the additive monoid N_0 to the multiplicative monoid Z, or rather **End**(Z). Since both answers are correct, it is a moot point whether one of them can be preferred to the other. Perhaps it is best to leave the question of preference to the metaphysicians. At least one can say that the investigation of this example has led to one important improvement in our point of view, if we had thought before that there could be only one logical answer to the question. Following up this suggestion of the muse, let us consider an arbitrary answer to the problem

$$1, 2, 4, 8, 16, \underline{\hspace{1cm}}.$$

An arbitrary answer means any old answer at all. Think of a number. It need not be a number between 1 and 100; it need not be less than a million; it need not be short enough to write down, even if one is allowed only one atom of any element at all for each letter. There is no restriction on the number one is allowed to write down as the answer; but to avoid unnecessary explanations we shall restrict it to be an integer. Just write this number on the line provided. (This line may be extended if it is not long enough; and if the number is too long to actually write during the probable lifetime of the reader, simply think of the number as written in.) Since I do not know what number you have written in, let us call it a_5. Since there is nothing special about the number 5, let us assume that to an arbitrary finite sequence

$$(a_0, a_1, \ldots, a_{n-1})$$

the reader has added one more number, making it

$$(a_0, a_1, \ldots, a_{n-1}, a_n).$$

Then the polynomial-generating formula that was mentioned earlier will produce a formula that will not only interpret the given terms $a_0, a_1, \ldots, a_{n-1}$ but will also include the new term added arbitrarily by the reader in the interpretation. This gives us the clear and simple rule that was sought, which would enable any fool to answer these little puzzles with a minimum of difficulty. The rule is this: you probably have a favorite number—most of us have one. If you do not know what your favorite number is, try to find out or decide somehow; perhaps an astrologer could help. Preferably, add 1 to this number; but if you are unable to perform this calculation, it does not matter greatly. Now take the number you have arrived at, and whenever you are given one of these questions involving guessing the next number in a sequence, use this number. (The addition of 1 to your favorite number is simply a device that makes it more difficult to determine your character defects by analyzing your number. No technique by which a person's character may be found out from his secret number is known to the author, but of course someone may some day invent such a technique.) For example, here is how you may answer the quiz we started with:

(1)   5, 4, 3, 2, <u>19</u>.
(2)   2, 4, 6, 8, <u>19</u>.
(3)   1, 3, 5, 7, <u>19</u>.
(4)   1, 2, 4, 8, 16, <u>19</u>.

There can be absolutely no question that the answer given here is correct, and that it tends to show a very high level of understanding of pattern-recognition. It must be admitted immediately that some of the people who evaluate these tests are so poorly equipped, patternwise, as to be unable to recognize the correctness of the answer given above. Of course, if the object is to guess the number chosen by the examiner as the one he intends to accept as correct, then a study of parapsychology is more relevant than a mathematical study of which answer is correct. Parapsychology is also called E.S.P. or mind-reading, and is a much more useful art than is mathematics. A good esper can pass an examination in any subject at all, so long as the examiner can also pass his own exam. Still, even E.S.P. is helpless if the examiner himself does not know the correct answer.

The reader may be assured that in writing the following it is not mere random whimsy that guides the author's unerring finger:

$$0, 0, 0, 0, 0, 0, 0, \ldots.$$

There is something very definitely in the writer's mind when he sets out that string of numbers. If this is regarded as to be completed by the addition of one more number, then that number may as well be 0, or 19. If the object, on the other hand, is to find out just what are the numbers (all of them) in the infinite sequence that I happen to have in mind, then even an accomplished mind-reader cannot tell you very much about the numbers. He can tell you something: since what I have in mind is that the term in the nth place is 0 if $n = 0$, 1, or 2; and 0 if $n \leq 3$ and if there exists no solution in integers to the equation

$$a^n + b^n = c^n$$

such that $abc \neq 0$; and 1 if $n \geq 3$ and there does—since all that is what I have in mind, he can tell you it. What he cannot tell you is whether all the terms of the sequence are 0, or some of them are 1. He cannot tell you that because the answer is unknown, not only to him and to me but to everyone. He has no minds to read about the answer in. That sort of question is a good one on any intelligence test: namely, the sort of question that has never been answered, but is known to have only one correct answer. At the opposite extreme from this is the question that is known to have all possible answers as correct answers; indeed it is strange that people not only get great enjoyment out of answering such questions, but that only intelligent people are able to answer them.

Since Andrew Wiles proved Fermat's Last Theorem, it is now known that the sequence based on $a^n + b^n = c^n$ does consist entirely of zeros. But there are plenty of other examples where that one came from. Since no one knows whether there is an odd perfect number (one that is the sum of its smaller divisors, as $6 = 1 + 2 + 3$ and $28 = 1 + 2 + 4 + 7 + 14$, but is odd), the sequence

$$a_n = \begin{cases} 0, & \text{if } 2n+1 \text{ not a perfect number} \\ 1, & \text{if } 2n+1 \text{ is a perfect number} \end{cases}$$

has the same status as the Fermat's Theorem sequence before Wiles's proof.

True to his goal of making mathematics difficult, Professor Linderholm does not show the easy way to get his polynomial

$$a_n = 1 + \tfrac{7}{12}n + \tfrac{11}{24}n^2 - \tfrac{1}{12}n^3 + \tfrac{1}{24}n^4$$

that gives 1, 2, 4, 8, 16, 31 for $n = 0, 1, 2, 3, 4, 5$. The easy way is to make a difference table,

n	a_n	Δa_n	$\Delta^2 a_n$	$\Delta^3 a_n$	$\Delta^4 a_n$
0	1				
		1			
1	2		1		
		2		1	
2	4		2		1
		4		2	
3	8		4		
		8			
4	16				

and use the elements in the first downward diagonal in Newton's Forward Difference Formula:

$$a_n = 1 + 1 \cdot \binom{n}{1} + 1 \cdot \binom{n}{2} + 1 \cdot \binom{n}{3} + 1 \cdot \binom{n}{4}$$
$$= 1 + n + \frac{n(n-1)}{2} + \frac{n(n-1)(n-2)}{6} + \frac{n(n-1)(n-2)(n-3)}{24},$$

from which his polynomial follows.

Applying the same method to 8, 75, 3, 9, we get

n	a_n	Δa_n	$\Delta^2 a_n$	$\Delta^3 a_n$
0	8			
		67		
1	75		-139	
		-72		217
2	3		78	
		6		
3	9			

so the cubic polynomial that gives these values is

$$a_n = 8 + 67n - 139\binom{n}{2} + 217\binom{n}{3}.$$

This can be used to find the next value, or the table can be extended:

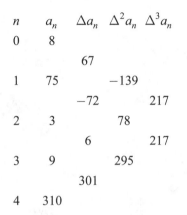

n	a_n	Δa_n	$\Delta^2 a_n$	$\Delta^3 a_n$
0	8			
		67		
1	75		-139	
		-72		217
2	3		78	
		6		217
3	9		295	
		301		
4	310			

So, the next number is 310. Or perhaps 17.

In response to a request for information about himself (in 1999), Professor Linderholm sent

Carl Linderholm has learned since he was discovered crawling on the floor that nobody is very certain why he entered this place. Mathematics is not known to answer such problems, but he engages in messing with it anyway. He also likes to travel, and his favorite poem in Irish is (in English) "Portrait of the Artist as an Abominable Snowman," by Gabriel Rosenstock. He is best known as the author of a book, *Mathematics Made Difficult*, which was completely ignored by the public at large. The title has been translated into Spanish, but not the rest of the book.

26
Mathematical Humor

It's not easy to be funny and mathematical at the same time. You can't make fun of algebra, or differential equations—there is nothing there that is absurd, or incongruous, to give an opening. There are no pretensions in mathematics to be punctured. So, about the best that can be done is parody, using the lofty and formal language of mathematics for something silly. The best known example is probably R. P. Boas's list of methods to catch a lion in the desert (which can be found, with several extensions, in *Lion Hunting & Other Mathematical Pursuits*, edited by Gerald Alexanderson and Dale Mugler, Mathematical Association of America, 1995). The fact that its original publication was in 1939 and that it is still quoted shows how rare good mathematical humor is.

Here is another example, not quite as old and not as well-known.

On the Set of Legs of a Horse
Marlow Sholander (1950)

In a recent issue of this *Journal* [1] there appeared the question "How many legs has a horse?", and the answer "Twelve; two in front, two behind, two on each side, and one in each corner." The enumerator overlooks the existence of bottom legs, outside legs, and so on. It is the purpose of this paper to give the correct answer to this fundamental problem. We prove that the number is infinite; more precisely, that the set of legs has the power of the continuum. This answer may come as a surprise to the casual observer but it was not unanticipated. There is some reason to believe that the originator of the word "equator" (from "equus" meaning "horse" and "ater" meaning "dark") was attempting to announce this discovery in the subtle fashion peculiar to his period. Again, Stephen Leacock [2] writes of someone who jumped on his horse and rode madly off in all directions. This, of course, would be impossible with a finite-legged horse. One authority on horseplay [3] has even made an attempt at a proof. He starts with the assumption that the number of legs, y, of a horse is a nondecreasing function of its age, x (the variable x is, of course, determined by the inspection of the horse's teeth). He then seeks to show that y is unbounded in any finite interval of x. However his proofs invariably begin "Given a horse ... " and he unfortunately overlooks that he has thus made x indeterminate since gift horses are not to be looked in the mouth.

We offer below several proofs of the fundamental result.[1] First let us note that the answer is not as unreasonable as it may first appear. Assume that the number of legs, n, of a horse is finite. From [1], we have $n > 4$. Even the most skeptical will grant this (n) is an odd number of legs for a horse. But, in contradiction, we have that n is even since each horse has a plane of symmetry which contains no legs. If an argument of a direct type is desired, the reader will recall that n is infinite if and only if the set of legs can be placed in one-one correspondence with a proper subset of itself. But if each leg corresponds to itself, we have such a correspondence—if the set of legs of a horse is improper, how does one account for the disappearance of the horse-blanket?

Perhaps the simplest rigorous proof can be made by using the well-known schachprinzip or "box principle".

Proof I. Assume that the number of legs, n, is finite. As the horse remains standing, place his legs in $n - 1$ boxes, so that at most one leg is found in each of the boxes. By the box principle, at least one leg is left over. But this is a contradiction since throughout the process each leg was left underneath.

Proof II. Suppose that there are some horses with a finite number and some with an infinite number of legs. This would clearly require the existence in every town of one each of two kinds of veterinarians, i.e., the existence of a paradox. Thus it is sufficient to show that at least one horse has an infinite number of legs. We prove more—that given any two horses, at least one has an infinite number of legs. Assume that all finite-legged horses have been colored red and all infinite-legged horses blue. Consider two horses. We may assume

---

[1] One remark is in order. Here and there a wording may seem ambiguous. In these cases, the reader is to interpret the words in their horse sense.

the first horse is red. If the second horse is blue, the theorem is proved. If the second horse is red, that's a horse of a different color. But in this case we have an immediate contradiction.

Proof III. We prove by induction that the number of legs, n, of a horse is greater than a positive integer m, for all m. By [1], this holds for $m = 1, 2, \ldots$, 11. Assume it holds for $m = k \geq 11$, i.e., assume the number of legs exceeds k. If the reader grants this, he will surely grant that n exceeds $k + 1$. If, on the other hand, he feels this is too too many, we have $n = k + 2$. Thus, in either case, the statement holds for $m = k + 1$.

Once we grant that a horse has an infinite number of legs we suspect the set is uncountable for the simple reason that people who have tried to count the set usually obtain 4 as an answer. This suspicion is confirmed in the following proof that the set of legs has the cardinal number of the continuum.

Proof IV. By definition, c, the cardinal number of the continuum, is the number of points in a lion. We wish to prove $n = c$. Let h be the number of points in a horse. Clearly $n \leq h$ and, from the case of a hungry lion and a small horse, we see also that $h \leq c$. It remains to be proved that $c \leq n$, i.e., that the points of a lion may be placed in one-many correspondence with the set of legs of a horse. But, trivially, we may have a one-many correspondence from a subset of the points of a lion to the set of legs of the horse. Regardless of which such correspondence we choose to consider, there is surely no point in a lion without a leg. Hence the correspondence considered is the one needed to complete the proof.

The author has as yet not had time to develop the innumerable consequences of this result. For one thing, it is clear that we have been misguided in shooting a horse when he breaks a leg since each leg has as its horsepower at most lim $1/n$. It would be of interest also to generalize the result $n \to \infty$ from horses to, say, the best cattle—those from which we get prime legs of beef. Those who make it a point to keep abreast of developments in foreign research inform the author that an Alexandrian geometer, Euclid, has already made progress in this direction. In closing, the author expresses regret that lack of space prevented his giving Proof V, perhaps the most interesting proof of all. This proof is based on an argument which makes use of the neigherhooves of a Horsedauf Space.

Bibliography

1. *Pi Mu Epsilon Journal*, 1 (1949) p. 24.
2. Stephen Leacock, Gertrude the Governess, or Simple Seventeen, *The Leacock Roundabout*, New York, 1946.

3. N. Baerbaki, Die Theorie und Anwendung des wiehernden Gelächters, Appendix
    II.

Note: N. Baerbaki is not to be confused with his distinguished French cousin N. Bourbaki. The former, in reply to a letter asking:

A. When his manuscript is to be printed,

B. If his last name is pronounced "barebacky",

C. What his first name is,

cabled, from his present address in eastern Germany, as answer (presumably only to the third question) the single word "Nichevo".

Marlow Sholander earned his Ph. D. degree at Brown University in 1949 and taught at Washington University, the Carnegie Institute of Technology, and Case Western Reserve University. He was a deadpan humorist and an author of verse, of which the following (from *Mathematics Magazine* 44 (1971) #3, 164) is a particularly nice example:

QED

"Now given this …", "If given that …",
Night after night the dreamer sat
And showed what he would do if he
Were shown some generosity.

Sources

1. Jean Dieudonné, *Mathematics—The Music of Reason*, translated by H. G. and J. C. Dales, Springer-Verlag, 1992, pp. 7–17. With kind permission of Springer Science and Business Media.

2. Morris Kline, *Mathematics and the Physical World*, Dover Publications, 1981, pp vii–ix, 464–475. With kind permission of Dover Publications.

3. Nathan Altshiller Court, *Mathematics in Fun and in Earnest*, Dial Press, 1958, pp. 96–107.

4. *Hansard* (House of Commons Daily Debates), 26 June 2003, vol. 407, part 117, columns 1260–1268.

5. Richard K. Guy, There are three times as many obtuse-angled triangles as there are acute-angled ones, *Mathematics Magazine* 66 #3 (1993), pp. 175–179.

6. David Hemenway, Why your classes are larger than "average", *Mathematics Magazine*, 55 #3 (1982), pp. 162–164.

   Wong Ngai Ying, Why is a restaurant's business worse in the owner's eyes than in the customers'?, *College Mathematics Journal*, 18 #4 (1987), 315–316.

7. Joseph Gallian, Assigning driver's license numbers, *Mathematics Magazine*, 64 #1 (1991), 13–22.

8. Richard K. Guy, The strong law of small numbers, *American Mathematical Monthly*, 95 #8 (1988), pp. 697–712.

9. Richard J. Trudeau, *The Non-Euclidean Revolution*, Birkhäuser, 1987, pp. 118–131. With kind permission of Springer Science and Business Media.

10. Patricia Cline Cohen, *A Calculating People: The Spread of Numeracy in Early America*, Routledge Publishing Inc., 1999, pp. 116–138. Used with permission of the author and the Copyright Clearance Center.

11. Paul R. Halmos, *I Want to be a Mathematician*, Mathematical Association of America, 1988, pp. 255–268.

12. *Encyclopædia Britannica*, A. Bell and C. Macfarquhar, Edinburgh, 1771, vol. 2, pp. 607–611, plate LXXXI.

13. David Eugene Smith, On the origin of certain typical problems, *American Mathematical Monthly*, 24 #2 (1917), pp. 64–71.

14. Lewis Carroll, *A Tangled Tale*, Dover, 1958, pp. 4–12, 84–89. (First edition 1885.)

15. John Aubrey, *Brief Lives*, The University of Michigan Press, 1957, pp. 229–233.

16. Jerome Cardan, *The Book of My Life*, translated from the Latin *De Vita Propria Liber* by Jean Stoner, Dover, 1962, pp. 19–25, 73–74, 97–118. Original edition, Dutton, 1930.

17. J. L. Synge, George Boole and the calculus of finite differences in *George Boole, A Miscellany*, pp. 4–18, Copyright 1969 J. L. Synge, by permission of Cork University Press, Youngline Industrial Estate, Pouladuff Rod, Togher, Cork Ireland.

18. Steven B. Smith, *The Great Mental Calculators*, Copyright 1983 by Columbia University Press, pp. 221–226, 289–298. Reprinted with permission of the publisher.

19. James Smith, *The Quadrature of the Circle*, Simpkin, Marshall, 1861, pp. vii–xxv.

20. Arthur E. Hallerberg, Indiana's squared circle, *Mathematics Magazine* 50 #3 (1977), pp. 136–140.

21. Edward Rothstein, *Emblems of Mind*, University of Chicago Press, 2006, pp. 135–140, 191–201.

22. G. L. Alexanderson and L. F. Klosinski, On the value of mathematics (books), *Mathematics Magazine* 55 #2 (1982), pp. 98–103.

23. Robert L. McCabe, Theodorus' irrationality proofs, *Mathematics Magazine*, 49 #4 (1976), 201–203.

24. R. J. Gillings, The so-called Euler-Diderot incident, *American Mathematical Monthly*, 61 #2 (1954), pp. 77–80.

25. Carl E. Linderholm, *Mathematics Made Difficult*, World Scientific, 1972, pp. 154–155, 89–99.

26. Marlow Sholander, On the set of legs of a horse, *Pi Mu Epsilon Journal*, 1 #3 (1950), pp. 103–106.

About the Editor

Underwood Dudley earned his B.S. and M.S. degrees from the Carnegie Institute of Technology and his doctorate (in number theory) from the University of Michigan. He taught briefly at the Ohio State University and then at DePauw University from 1967–2004. Woody has written six books and many papers, reviews, and commentaries. He has served in many editing positions, including editor of *The Pi Mu Epsilon Journal*, 1993–96 and *The College Mathematics Journal*, 1999–2003. He is widely known and admired for his speaking ability—especially his ability to find humor in mathematics. He was the PME J. Sutherland Frame lecturer in 1992 and the MAA Pólya lecturer in 1995–96. Woody's contributions to mathematics have earned him many awards, including the Trevor Evans award, from the MAA in 1996, the Distinguished Service Award, from the Indiana Section of the MAA in 2000, and the Meritorious Service Award, from the MAA in 2004.